HOW OUTER SPACE MADE AMERICA

For Catherine and Edward

How Outer Space Made America
Geography, Organization and the Cosmic Sublime

DANIEL SAGE
Loughborough University, UK

Routledge
Taylor & Francis Group

LONDON AND NEW YORK

First published 2014 by Ashgate Publishing

Published 2016 by Routledge
2 Park Square, Milton Park, Abingdon, Oxon OX14 4RN
711 Third Avenue, New York, NY 10017, USA

First issued in paperback 2018

Routledge is an imprint of the Taylor & Francis Group, an informa business

British Library Cataloguing in Publication Data
A catalogue record for this book is available from the British Library

The Library of Congress has cataloged the printed edition as follows:

Sage, Daniel, 1980–
    How Outer Space Made America: Geography, Organization and the Cosmic Sublime / by Daniel Sage.
        pages   cm
    Includes bibliographical references and index.
    1. Outer space – Exploration – United States – History. 2. Outer space – Exploration – Social aspects – United States. 3. Astronautics – United States – History.
    4. Astronautics – Social aspects – United States. 5. National characteristics, American.
    I. Title.
    TL789.8.U5S24 2014
    629.4'10973–dc23   2014011764

ISBN 13: 978-1-138-54668-4 (pbk)
ISBN 13: 978-1-4724-2366-5 (hbk)

Contents

List of Figures

Acknowledgments

Across several years, and two countries, numerous individuals have offered their support, inspiration and guidance, in the development of this book. I am especially grateful to the following: Anne Collins-Goodyear, Deborah Dixon, Marcus Doel, Colin Fries, Daniel Gruembaum, Debra Land, Margaret Weitekamp and Michael Woods, as well as Katy Crossan, my editor as Ashgate. I am also indebted to the support for social science research in three Space organizations: Kennedy Space Center Visitor Complex, NASA's History Office and the National Air and Space Museum. And finally, I must thank those closest to me for their patience and encouragement—my wife, Catherine, and my son, Edward.

Introducing a Geography of Outer Space

Those passionate about outer space have long been in awe of its apparent 'spacelessness,' outer space appears unbounded, infinite, sublime. When we see or think through Space[1], whether by looking at images produced by a powerful space telescope or enjoying a science-fiction film, we can journey in an instant to the most distant reaches of the universe, and simultaneously billions of years back in time, or into a barely imaginable future, far beyond the possibility of human life. If we think about the origins or death of the universe we can consider a primordial cosmic singularity before the Big Bang where time and space were absent, or an eventual cosmic Heat Death, when all energy and matter is effectively 'dead.' Cosmology offers thought an intensely transcendental experience. But this experience is not totally unique to modern humans. For millennia humans have, across various religions, imputed the cosmos with a capacity to transcend the here and now of human experience (Dickens and Ormrod, 2007; Sagan, 1994).

And so, if Space is spaceless, then how does it relate to geography? After all, surely geography is a discipline concerned with the 'groundedness' of mappable points, territories, relations and locales: geography seems to be an anathema to those thoughts, individuals, objects or machines that are pulled into this sublimely mysterious realm.[2] In other words, Space, and the transcendental experiences it offers, seemingly falls outside of geography's disciplined passion to describe and explain concrete spaces and places that are mappable, spaces which matter, which relate to *us*, rather than render us absent, or Other to ourselves. Yet, of course, Space has always enabled new opportunities to look back at Earth, to look back at ourselves, as much as out and beyond us. Just as ancient peoples looked to the heavens to interpret their own position on Earth (Cosgrove, 2001; Dickens and Ormrod, 2007; Livingstone, 1993; Sagan, 1994), we still yearn to understand ourselves from beyond ourselves, to determine, to map, and (re)place *us* in the vast cosmos, or simply from Earth orbit (Macdonald, 2007; Parks, 2005). Carl Sagan's successful request to NASA in 1990 to turn Voyage 1 spacecraft's camera to photograph the Earth as a 'pale blue dot' from a distance of 3.7 billion miles, as it sped out of the Solar System at 40,000 miles per hour, vividly illustrates this yearning to place ourselves in Space (Sagan, 1994: 1-2). As Voyager 1's camera

---

1   The capitalized word 'Space' is also used to refer to outer space, or the cosmos; capitalization is intended to distinguish it from abstract concepts of space.

2   The terrestrial grounding of geography is a relative recent invention. Prior to the institutionalization of the discipline in the early twentieth century, there was much greater interplay between geography, as factual Earth description, and cosmography, as a quasi-mythical ordering of the universe (Cosgrove, 2001; Livingstone, 1993).

turned and the images travelled for over five hours to reach Earth, we placed ourselves in the Universe, discovering our space, our time, at the very moment we appeared lost:

> It has been said that astronomy is a humbling and character-building experience. There is perhaps no better demonstration of the folly of human conceits than this distant image of our tiny world. To me, it underscores our responsibility to deal more kindly with one another, and to preserve and cherish the pale blue dot, the only home we've ever known (p7).

Lisa Parks, in her book, *Cultures in Orbit*, explains how this 'Janus-faced vision,' this simultaneous looking beyond and back at ourselves from Space, does not actually require a camera to be turned back towards Earth—even the Hubble Space Telescope can find us again in the cosmic depths: 'So long as there persists an unwillingness to imagine and accept difference in astronomical fields, Hubble will continue to show us only what we can bear to look at, whether the catastrophic disaster of another planet, the humanlike productivities of stellar matter, or Earth's central position in the cosmos' (Parks, 2005: 158-9). Perhaps the most well cited incarnation of this cosmic humanization is found in T.S. Elliot's poem *Little Gilding*: 'We shall not cease from exploration, And the end of all our exploring, Will be to arrive where we started, And know the place for the first time' (Eliot, 1944: 43).

By acknowledging these processes of cosmic 'humanization' (Dickens and Ormrod, 2007), we open up analytical purchase on Space for the social sciences and humanities, not least for geography: Space is not *only* a disorientatingly spaceless and timeless abyss, somewhere where geography with its passion for mapping socio-spatial points, surfaces, hierarchies and relations cannot travel; rather through Space we can space and place ourselves in countless new ways, albeit in pursuit of well-established interests and agendas. Accordingly, in recent years, geographers have been central figures in burgeoning *critical* debate about the humanization of space travel across the social sciences, spanning Sociology (Dickens and Ormrod, 2007), History (Geppert, 2012a; Hersch, 2012; Macauley, 2012), Organization Studies (Parker, 2009a), Science and Technology Studies (Redfield, 2002), Film and Media Studies (Parks, 2005) and International Relations (Peoples, 2009). Within this wider effort, geographers have, thus far, provided critical interrogations of: NASA's Earth photographs from the Apollo program (Cosgrove, 2001), nineteenth-century Martian cartographies (Lane, 2011), contemporary Martian robotic exploration (Dittmer, 2007), military surveillance and weaponization (Gregory, 2011; Macdonald, 2007), Cold War Rocketry (Macdonald, 2008), British astronomy (Dunnett, 2012), space law (Collis, 2009) and space tourism and regional economic development (Beery, 2012). As a result of such analyses, our capacity to view ourselves both from, and in, Space, has been connected to a multitude of more or less interwoven human projects: ethno-centrism (Dittmer, 2007; Macdonald, 2007; 2008), nationalistic imperialism (Lane, 2011; Macdonald, 2007; 2008), globalization (Cosgrove, 2001), environmentalism (Cosgrove, 2001), entrepreneurialism (Beery, 2012), Darwinian colonialism (Dunnet, 2012; Dittmer, 2007; Lane, 2011), scientific rationalism (Lane, 2011), and neoliberalism (Collis, 2009; Macdonald, 2007). As Macdonald (2007) puts it:

> Space is already being produced in and through Earthly regimes of power that undoubtedly threaten social justice and democracy. A critical geography of space, then, is not some far-fetched or indulgent distraction from the 'real world'; rather as critical geographers we need to think about the contest for outer space as being constitutive of numerous familiar operations (p611).

Fraser Macdonald's concern here about becoming lost in the sublime Otherness of space is well-established. In 1963 the political theorist Hannah Arendt criticized space exploration as a transcendental quest to find an Archimedean vantage point, to survey the Earth, which would allow Science to detach itself from humanity, as the Earth, and its political struggles, become speciously reduced to a preordained biological process (Arendt, 1963). More recently, the sociologists, Peter Dickens and James Ormrod (2007), in their book *Cosmic Society*, argue, via a blend of Neo-Marxism and psychoanalysis, that all daydreaming about transcendence of the Self in Space operates as part of a childish narcissistic fantasy—a 'God Complex'—undertaken by a, usually Western, male, and rich, cosmic elite who aggrandize their sense of Self in dreams of being 'intermediaries,' or 'demi-Gods,' in a 'New Chain of Being.' Similarly, introducing the edited collection *Imagining Outer Space: European Astroculture in the Twentieth Century*, the historian, Alexander Geppert, proposes that 'Space enthusiasm and terrestrial geocentrism are two faces of the space coin' (Geppert, 2012b: 5).

These humanistic ideas of the sublime owe much to the philosophy of Immanuel Kant. The relationship of the Kantian mathematical sublime[3] to the cosmos is usefully explained in this passage by the French philosopher Jean-François Lyotard:

> The cosmos is certainly not an object of experience, it escapes our sensory intuition, it is the absolutely vast. But our reason can still form the Idea of it. Our capacity for feeling and even for imagining it is cause for sorrow, but this unfortunate finitude is also that which inform us of the infinite capacity of our

---

3 Kant actually describes two concepts of the sublime: 'mathematical' and 'dynamic.' Throughout this study reference will be made to both of these concepts; however, while they are different (the latter referring to an experience of infinity, such as the cosmos or God, the former the contemplation of physical terror, such as a volcano or avalanche) in both cases our faculty of Reason prevails and is aggrandized (either by being able to grasp infinity, if only partially, or the intellect being immune to physical danger). For a discussion see Nye (1994: 78-)

reason ... This goodess is free reason in us, the thought of the absolutely vast
(Lyotard, 1998: 208–9).

Lyotard himself challenges this humanistic version of the sublime by acknowledging
how modern astrophysics supplants the negative sublime movement between
painful finite experiences of the cosmos and joyful Ideas of the vastness of outer
space, with a new positive sublime wherein an infinity of possible technological
set-ups and scientific theorems serve to limit our everyday experience (and Idea
of Self): 'The sublime of immanence replaces the sublime of transcendence' (p.
229). For Lyotard we now only feel absurd humour, not pain or pleasure at this
new feeling of the sublime. More hopefully, David Bell and Martin Parker suggest
in their introduction to the edited volume, *Space Travel and Culture*, that 'any
form of alterity, whether expressed in terms of great distances in time and space, or
objects of power and size, might do enough to displace the observer from common
sense, and allow them to see the world differently' (Bell and Parker, 2009b: 3).

In the wake of such nuanced ruminations on what might be termed the
'cosmic sublime,' my purpose in this book is to take seriously such transcendental
engagements with Space, not to deride them as childish narcissistic fantasies, but
rather address them as sublimely Other ways of producing ourselves, our futures,
on Earth and beyond. Often these thoughts and practices are hubristic, sometimes
hopeful, absurd, and occasionally boring. Rather than pursue a philosophical
treatise on the space age sublime, across the following chapters I argue that
cosmic thoughts which appear spaceless, timeless, and transcendental, have long
been connected to all manner of seemingly more grounded geographies: nations,
places, locales, relations, organizations, landscapes, museums and popular
cultures. Instead of regarding these more than representable geographies of Space
as something Other to more Earthly concerns (as in Geppert, 2012b: 5; Parker,
2009a: 330), I seek to show how such sublime geographies are fundamental to
the spacing and placing of, in particular, one nation: the United States of America.
Before discussing in more detail the reasons why I have chosen to focus on US
space travel, I will explain how the thoughts developed here are, in part, inspired
by, framed within, and extend, ways of thinking with space within Deleuzo-
Guattarian philosophy.

Thinking with Cosmic Space

Within Gilles Deleuze and Felix Guattari's *A Thousand Plateaus* (1987) we are
provided with an explicitly spatial, if mysterious sounding, philosophy: concepts
such as 'lines of flight,' 'nomadology,' 'strata,' 'territories,' 'mapping,' 'surveying'
as well as 'territorialization' and 're/deterritorialization' abound. Articulated
across these concepts are two specific modes of thinking *with* space: 'smooth'
and 'striated' space. The former resembles the desert where the rhythms and
movements of nomadic hunters, prey, sand, wind, water and landscapes emerge

in their playful interplay; the latter corresponds to the oasis town, on the desert's edge, populated by more sedentary people and animals whose movements follow well-worn tracks around fixed water sources, homes, markets and agricultural land. Importantly, these modes of space are not conceptualized in opposition to one another; rather they are closely related: 'the two spaces in fact exist only in a mixture ... one organizes even the desert; in the second the desert gains and grows; and the two can happen simultaneously' (p. 474–5). Deleuze and Guattari (1987) are not proposing here some sort of social geomorphology, but rather a geo-philosophy (Deleuze and Guattari, 1994): a 'project that entails thinking [with] earth, ground, land and territory' (Bonta and Protevi, 2006: 8).

As Buchanan (2005: 5) explains, in conceptualizing these two modes of thinking with space, Deleuze and Guattari oppose the subject-object model of space, popularized by scholars such as Henri Lefevbre, across and beyond geography. Instead of space being conceived of in terms of containers of 'reality-representation-subject', Deleuze and Guattari suggest that thinking *with*, rather than *about*, space implies no subjective centre or external margins; all stable wholes, including conscious thought, is an effect of spatialized processes of smoothing (of 'deterritorilizaiton') and striation (of 're/territorialization'). Thus, our brains, bodies, houses, gardens, towns, organizations, States and Earth, emerge through 'molar' striations which are effects of organizations of matter, energy and thought, induced by concurrent 'deterritorializing' 'molecular' 'lines of flight,' or 'becomings,' whether flows of wind, sea, noise, art, mathematics, philosophy, animals, disease, migration, or global capitalism (Bonta and Protevi, 2006; DeLanda, 2006). And so, for example, factories are lost to the accelerated flows of money of capitalism, just as bodily tissues are lost to disease. In turn, such deterritorializations prompt reterritorializations: attempts to substitute something, as a token, for that which is lost. So we build modern apartments to replace lost factories (Buchanan, 2005: 31) or replace lost organs with artificial ones: space is reconstituted, organized again, but not as before. How then can these ways of thinking *with* space relate to Space?

Deleuze and Guattari (1987) only mention space travel once in *A Thousand Plateaus*. With reference to the 1960s space race they describe how the US sought to effectuate the 'deterritorialization' of capital "to the moon"' (p. 455) (and presumably beyond); however 'the USSR, which conceived of extraterrestrial space as a belt that should circle the earth taken as the "object"', (p. 455), aborted their lunar program; thus the Earth was produced as an a object for reterritorialization of capital, and further striation. The important point here, that Deleuze and Guattari (1987) make all too briefly, is that outer space, just like the Earth, is *simultaneously* composed of a triad of interwoven processes of deterritorialization and reterritorialization. Space is replete with emergent lines of flight, where coherent subjects and objects become imperceptible from each other in emergent, indeterminate movements (Deleuze and Guattari, 1987: 480): these flows of thought, matter and energy are excessive to our attempts to render them visible, whether the initial singularity, moments before the Big Bang, when space and time (and all life) did not (yet) exist, to the celestial transcendence of an omnipotent and omnipresent God of Judeo-Christian cosmologies. But Space is also occupied by 'molar' reterritorializing practices that organize it by drawing lines between fixed points (whether the mapping of Mars for colonization—Dittmer, 2007; Lane, 2011; the monitoring civilians with GPS on Earth—Macdonald, 2007, or the legal mapping of Earth orbit—Collis, 2009).

Rather regrettably, Deleuze and Guattari (1987) do not discuss space exploration in more detail. Instead they develop a series of other models of smooth/striated space. Arguably the most related to outer space is their maritime model: the sea is a 'smooth space par excellence' (Deleuze and Guattari, 1987: 479) as despite centuries of ocean mapping, vessels, like submarines, can flow through it to evade being mapped, acting like nomads drifting in the desert; however, submarines move with the 'purpose of controlling striated space (i.e. the land) again more completely' (p. 477). Indeed, as Deleuze and Guattari (1987) explain 'smooth spaces are not in themselves liberatory' (p. 500)—they can and do service imperial States to effect more striation, more control. Likewise, military surveillance of civilians and enemy combatants now routinely occurs from Earth orbit (Macdonald, 2007).

Deleuze and Guattari's (1987) maritime model of smooth space, and their brief mention of the Space Race, suggests that the significance of the cosmos is that it offers an un-representable smooth space from which to effect more Earthly striations; however, this move does not really fully permit us to take seriously the spacelessness of Space, and specifically its sublime character: if Space is *only* a celestial sea through which to effect terrestrial striations more forcibly, perhaps via a military satellite or submarine launched ballistic missile, then we soon arrive back again at a geography of mappable points, lines and relations. Perhaps the cosmos (although maybe not Earth's orbital zone—Collis, 2009) is a smooth space *par excellence*: energy, matter and thought can be rendered transcendent in the cosmos, operating beyond striations of space and time, and beyond ourselves, as we consider the initial singularity of the Big Bang, or narratives of spiritual transcendence. The cosmos appears an absolute threshold of deterritorialization, of smoothing. Yet following Deleuze and Guattari's geo-philosophy even 'Absolute deterritorialization does not take place without reterritorialization' (Deleuze and Guattari, 1994: 101).

How might the cosmos be reterritorialized? Deleuze and Guattari (1994) offer a clue in their last joint work, *What is Philosophy?*: 'In imperial states deterritorialization takes place through transcendence: it tends to develop vertically from on high, according to a celestial component of the earth. The territory has become desert earth, but a celestial stranger arrives to re-establish the territory to reterritorialize the earth' (Deleuze and Guattari, 1994: 86).

What imperial celestial stranger could be substituted for such a complete deterritorialization of the boundaries and relations between our Self, our striated spaces, our Earth? And how might it attempt to reterritorialize the cosmos? If so, to what end? And, indeed, what happens when this reterritorialization falls apart;

after all, following Delezue and Guattari (1987; 1994), no reterritoralization is final. I will argue across the following chapters that some salient answers to these questions can be found in the history of the United States of America, its culture, politics, religion and technology, and especially its space program.

The Transcendental State

Hereafter the analytical focus of this study shifts to the United States of America. In Deleuzo-Guattarian parlance, my intention here is to argue that *some* significant de/reterritorializaitons of the cosmos emerge amid relations between space exploration and American art, popular culture, museum culture, religion, technology, society and politics. In following this cosmic assemblage, manifest across different disciplinary objects of analysis, my analysis journeys across the disciplinary divides between Human Geography, Organizational Studies, Museum Studies, Art History, Cultural Studies, International Relations and Continental Philosophy. Tracing these flows brings to the fore formations, and deformations, of an image of America as a 'transcendental state.' I employ this concept as an analytical convenience for a diverse but overlapping set of discursive and material practices through which America assumed, assumes, and may yet still assume, an exceptional ordering of cosmic space and time. This ordering of space and time is highly hubristic and exclusionary, but can also appear absurd, hopeful, even boring: it operates to reproduce and serve a narrow set of social interests and groups, often with quite pernicious effects; yet it also induces opportunities to think through American society, culture and politics in ways that might be termed progressive, even, in a sense, liberatory.

In undertaking this study I do not intend to extend a Deleuzo-Guattarian geo-philosophy into the cosmos. And indeed a great deal of what follows will be only intermittently connected to their philosophy. But their insistence on the co-dependency of de/reterritorialization does frame my analysis. My study is principally an analysis of how American geopolitical power has been projected, legitimized and transformed, through space exploration. Geopolitical power, or simply 'geopower,' can be understood through the geographical sub-field of critical geopolitics. The raison d'être of this field is to ' ... disturb the innocence of geography and politicize the writing of global space ... to resist the exercise of geopower by those centers of modern authority who wish only to make the world in the image of their maps' (O Tuathail, 1996: 20). I consider seldom discussed connections between the writing of global space and cosmic space, and in so doing also challenge the tendency, often apparent within critical geopolitics approaches to focus analytically on (often visual/representational) practices of striation, over smooth deterritorializations (for more on this see Jones and Sage, 2010; Macdonald, 2006; O Tuathail, 2000: 390; Thrift, 2000).

The narrative I develop of American space exploration is not *simply* one of military surveillance or weaponization (Gregory, 2011; Macdonald, 2007), geostrategy (Dolman, 2001), technocratic management (Parker, 2009a; McDougall, 1985), ideological rivalry (Burrows, 1998, McCurdy, 1997), economic empires (Dickens and Ormrod, 2007) or ballistic missile defence (Peoples, 2009); rather it also involves a set of both more familiar and more fantasmic practices, responses to the un-representable cosmos, found across art, religion, popular culture, museum culture, work, political speeches and journalism. While these practices are not wholly removed from other stories of American space exploration, they do challenge the belief that the inception and development of American space exploration is at heart a tale of Cold war paranoia, technocracy and economic imperialism. I find much of value in such accounts, but I cannot help but feel that one question remains insufficiently unanswered: how if space exploration is so exciting, so awe-inspiring, so sublime, can it also be so imperialistic, so destructive, even so boring? In short, how can we reconcile, if only partially, the wonder of space with its drearily predictable acquiescence with American geopower. Dickens and Ormrod (2007) suggest one answer—it is merely a childish narcissistic escape fantasy for (usually) rich, white, men. But if this is all that can be said then the awe of space travel hardly plays any active role in American geopower; it simply acts to conceal a web of more insidious interests (Parker, 2009a). I will argue here that far from merely operating as a false ideology, an aesthetic of sublime transcendence, was, and is, and may yet be, far more central to the incubation of American geopower, and perhaps not just in Space, than is usually assumed, especially, but not exclusively, with reference to the activities of NASA and its human space flight program. Thus any geography of Space, must take its affective qualities, its spacelessness, as well as its spaces seriously. Before proceeding to specify the format of this book, I will elaborate the rationale surrounding my decision to focus upon American space exploration with reference to its terrestrial Other(s): postcolonial critique.

Making Space for the Other

Geographers (Cosgrove, 2001; Dittmer, 2007; Macdonald, 2007; 2008) and others (Dickens and Ormrod, 2007; Redfield, 2002), have repeatedly shown that space exploration remains a hugely ethnocentric, neo-colonial affair: only a small handful of nationalities have demonstrated a capacity to send people into Space without foreign support (currently the US, Russia and China), and few possess the finances, technology and skills capable of shaping the development of Space, let alone on the scale of the US. Redfield (2002) describes Space as part of a colonial sensibility:

> In the aftermath of the 20th century, advocates of space exploration constitute perhaps the last unabashed enthusiasts of imperialism, cheerfully describing conquest, settlement and expansion, and hesitating not a whit before employing the term 'colony.' Theirs is a Columbus of exploration, nation building and risk

taking, not of invasion, domination and genocide. History is cleansed above the planet (p. 797).

It is difficult to contest Redfield's postcolonial critique. Moreover, as Macdonald (2007: 607) observes in the context of US space exploration, far too often talk of transcendental humanism operates as a thin veil for the colonial exploitation of Space by imperial powers and elites, whether to function as a platform for new weapons and surveillance technologies, provide a source of new mineral resources, or instil technocratic government (Dickens and Ormrod, 2007; Parker, 2009a). Hence in telling a story of American space exploration, one which takes seriously notions of cosmic transcendence, is it not also possible I may inadvertently be legitimizing this neo-colonial adventure?

In response to such postcolonial critique, it is first important to note that notions of celestial transcendence are not restricted to Western cosmic elites (contra Dickens and Ormrod, 2007). The anthropologist, Duane Hamacher (2011) describes how eternal creation cosmologies that order social relations in Australian Aboriginal clans (the 'Dreamtime'), are often derived from on-going astronomical observations: 'Celestial names and stories are accounted to the various stars and celestial objects, with many stars serving as mnemonic devices or incorporated into class relationships and marriage classes' (p. 3). Notions of celestial transcendence also appear in other cosmologies, including the transcendental pantheism of Hinduism and Taoism, as well as the transcendental monotheism of Judaism, Christianity and Islam. Even popular television documentaries, such as the BBC/ Discovery Challenge/Science Challenge production, *Wonders of the Universe*, now present stories of cosmic singularities, Dark Energy, black holes and cosmic decay, inviting our thoughts into the infinite cosmic void, de-centring ourselves and the Earth. These varied transcendental cosmologies may require access to religion and popular space science, whether through religious worship, education, books, satellite television channels or web forums; but they are certainly not restricted to the Western, rich, male, elite of narcissistic cosmic daydreamers, described by Dickens and Ormrod (2007).

Clearly we can make a distinction between transcendental cosmologies and active participation in space travel; yet, at the very least, it seems difficult to imagine that we will be enthralled by the latter without at least considering our orientation towards the former. Parks (2005), for example, describes an aboriginal documentary called *Satellite Dreaming*, wherein the orbital path of a television satellite (used to transmit into the Australian outback) is connected to the cosmological journeys of the 'dreamtime'. Thus, 'Aboriginal satellite dreamings also challenge critical assumptions that satellite television, and cosmic transcendence, works *only* as an agent of Western cultural imperialism and neo-colonial control' (Parks, 2005: 73). By contrast, within the US, attempts have been made to capture cosmic myths by the State; wherein they have been more or less intentionally put to work for imperialistic ends, namely to foster American geopower. What is more, and as will be argued in Chapter 2, aspects of Judeo-Christian transcendental cosmology are particularly apposite to this task (in comparison to the immanent transcendence of the aboriginal dreamtime or Hindu pantheism).

Unsurprisingly, the transcendental orderings of space and time discussed here in relation to America have parallels to those in other (predominately) Judeo-Christian cultures, notably Europe (Geppert, 2012b) and the USSR; however there are important differences. In the case of Europe, Geppert (2012b) suggests we view the emergence of European astroculture as 'cosmic provincialism' *vis-à-vis* US-USSR Cold War hegemony. Imperialistic myths of space colonization would appear far harder to maintain without significant investment in a European human spaceflight program. The USSR (and post-Soviet Russia) offers a more compelling analogue. Not least because, as Rogatchevski (2011) explains, ideas of 'collectivism, planning and patriotism, frequently associated almost exclusively with Socialist realism are not exactly uncommon' (p. 253), within American space exploration itself (see also Parker, 2009a). Rogatchevski (2011: 260) further suggests that narratives of spiritual transcendence in the cosmos actually appear more dominant in current post-Soviet space culture, than in the West. Despite such parallels, we must bear in mind various constraints on the cultivation of a Soviet 'transcendental state': the absence of strong cultural, political and religious associations between nationalism, exploration and landscape (as present in the US—Stephanson, 1995); state attempts in the USSR to eliminate religion, including within the Space program (Smolkin-Rothrock, 2011); far higher poverty, repression and violence within the USSR than the US (serving at least to defer if not dispel socialist utopias—Smolkin-Rothrock, 2011: 72); and various early Soviet space failures (notably: the deaths of Chief Designer, Sergei Korolev in 1966, and the first human space traveller, Yuri Gagarin, in 1968, as well as the loss of the Space Race to the Moon in 1969).

American space exploration, especially human space exploration, can thus rightly be described as a more historically imperialistic adventure than that of the USSR/Russia, and certainly more so than Europe; yet it is too simplistic to say that notions of sublime cosmic transcendence *per se* can either account for, or conceal away, this imperialistic orientation. After all, very similar transcendental narratives appear in other ex-colonized and ex-colonizing cultures. Rather, what must be considered is how a very specific notion of transcendence, of the sublime, was mobilized in the US before and during the inception of space travel, how it was sustained or transformed, and thus how Space became entangled with a sense of what America was, is, and could yet become. These questions are the explicit focus of this study. More implicitly perhaps, by empirically interrogating disparate sites and times where these de/reterritorializing processes of the cosmos play out, and refusing to order them into a linear history of progress, the content and structure of this book modestly attempts to produces some lines of flight through which American geopower might be denaturalized. I will now outline the nine chapters that constitute this book.

Summary of Chapters

In Chapter 1 I discuss the historical development of an image of America as a transcendental state across Judeo-Christian religion, American politics and literature and into sublime landscape artwork. This image describes America as the Promised Land, a people set apart, where American frontier landscapes become evidence of America's position on humanity's spatial, temporal and moral vanguard. Pre-empting tensions revisited in subsequent chapters, I will show how this quixotic image was always as unstable as it was seductive.

In Chapter 2 I start examining how cultural imaginaries of the American transcendental state were first extended, and then developed, into Space. In this chapter I pay particular attention to the career of the once influential, but now largely forgotten, American space artist, Charles Bonestell, and his relationship with early proponents of spaceflight in the US, notably the former Nazi rocket engineer, Werner von Braun. Piecing together contemporaneous media and space policy documents, I discuss how key events in the history of American space exploration, such as the launch of Sputnik, the founding of NASA, and the election of President Kennedy, were all partly framed as a response to the image of America as the transcendental state, with significant consequences for American politics and culture.

Chapter 3 focusses on decision by President John F. Kennedy in 1961 to pursue a human landing on the moon. While the Moon landings possessed undoubted value to science and engineering, the primary impetus for this mission was ideological. Yet what is perhaps less readily understood is how the symbolic potential of the Moon mission was bound up with the sublime aesthetics of the transcendental state. In this chapter I evaluate the significance of this prominent, yet fragile, intersection of politics, landscape and mythology.

Turing towards the organization of NASA, Chapter 4 draws upon a range of historical documents and accounts to examine how NASA functioned as an idealized technocratic organization, where a blend of technological determinism and instrumentalism served to evidence the image of transcendental state. However, as I show, the material organization of NASA, as the epitome of a technocratic organization composed of spaces of control, calculation and efficiency, was constantly confounded by both the mythological image of transcendence through which it was founded, and the unruly technologies brought into play to render space exploration possible.

Chapter 5 addresses how NASA not only prescribed that exceptional human bodies and machines should interact under the banners of American progress and success, but which expert bodies were more inevitably more desirable in this technocratic future. As these bodily prescriptions were rationalized under managerialist discourses of efficiency and rationality, alongside transcendental myths of exceptionalism and destiny, NASA became a significant site through which American modernity was gendered in hegemonically masculine terms.

In the wake of mounting civil rights demonstrations, dwindling public interest in Apollo, President Nixon's critique of 'big' government, in Chapter 6 I consider how NASA responded to challenges to its technocratic mandate and mythological inception. Focusing on NASA during the administrations of Nixon, Ford and Carter, I explore how those leading NASA sought to negotiate the Space Transportation System (or Shuttle) to ensure an American lead in space exploration. Shadowing the rise of neoliberalism in American politics, the Shuttle would also provide an opportunity for NASA to remodel itself as profit-driven technocracy, exemplifying equally nationalistic notions of American opportunity, ingenuity and prosperity.

Returning back to the role of Space in American public imaginations, in Chapter 7 I focus upon how the image of America as a transcendental state is organized and disorganized through particular sites of memory: the Kennedy Space Centre Visitor Complex (KSCVC) and the National Air and Space Museum (NASM). Through auto-ethnographies of these sites, I discuss how each of these sites struggles to construct a progressive image of America on the vanguard of cosmic space-time, offering fleeting opportunities for transformation.

In Chapter 8 I draw upon theories of affect to consider how the image of America as the transcendental state was both challenged, and rehabilitated, across two traumas experienced by NASA: the losses of the Apollo 1 spacecraft and space shuttle Challenger. These two events shed light on the confident feelings and hopeful affects that sustain the image of America as *the* transcendental state, as well as some of their deleterious consequences.

By way of conclusion, in Chapter 9 I re-connect my conceptualization of the American transcendental state back to work within critical geopolitics and human geography more generally; this enable me to elaborate upon the contribution of this study to theories concerning the operation of American geo-power, alongside possibilities of geopolitical resistance and transformation to such imperialistic cosmographical imaginations.

Chapter 1
America as Transcendental[1]

A Land Set Apart and Beyond

In 1997, the historian Walter McDougall explained how American identity is orientated around the history of Puritan pilgrimage, whereby 'a people dedicated to liberty based on faith, who had begun history over again in a New World, might *confidently assume*, a future free of limits imposed by man' (p77; original emphasis). McDougall (1997) suggests a universal sense of freedom as being part and parcel of what it is to be American: a future without limits is *the* distinguishing trait of American identity. McDougall (1997) argues the voyage of Puritans to the New World was read as an interwoven material and spiritual journey. In the first instance, America offered, at least potentially, more land and resources than Europe; and spiritually, America offered a land of religious freedom, a story testified to by the many Pilgrim voyages from the seventeenth century avoiding religious discrimination in Europe (Agnew, 2005: 96; Dijkink, 2006: 202). So that, as Ricard (1999) puts it, for these Puritan colonizers the 'unifying element was the cult of freedom … ' (p15). Contained throughout the early stories of the American colonies, especially those of the New England Puritans, was a notion of spatio-temporal difference written, rather paradoxically perhaps, through transcendence—between the restraints of England and the freedoms of the New World (Noble, 2002; Stephanson, 1995). The gradual elaboration and refinement of these embryonic notions of New World self-identity within American cultural identities is the subject of this chapter.

O'Tuathail (1996) suggests that appeals to the transcendental, somewhat counter-intuitively, often do not refer to the displacement of geographical difference by the universal, but rather offer a powerful means of universalizing a partial vision of spatio-temporal difference. The passion for transcendentalism inscribes, as much as it effaces, geographical differences, between different peoples and lands. The Puritan technique for the inscription of geographical difference through transcendence was not the God-like perspective of Enlightenment reason, favoured by Old World imperialists 'objectively' mapping, claiming, and dividing, colonial possessions (O'Tuathail, 1996: 81); rather it was a far less secular notion: the proclaimed spiritual and material freedoms of the New World confirm upon its population a unique messianic destiny to lead humanity towards salvation. In time

---

1   Material in this chapter was first published in 'Framing Space: A Popular Geopolitics of American Manifest Destiny in Outer Space,' Sage, D, (2008), *Geopolitics* (Routledge), reprinted by permission of the publisher (Taylor & Francis Ltd, http://www.tandf.co.uk/journals).

this transcendentalism became articulated through various territorial mythologies, wherein America was described as the 'Promised Land' or the 'New Israel' (Noble, 2002: 4; Wallace, 2006). And so spiritually, spatially and temporally the New World appeared set apart—a place that for many would yet be 'an exemplary state separate from the corrupt and fallen [Old] world' (Stephenson, 1995: pxii) where history would appear to have 'begun ... over again' (McDougall, 1997: 77). Ultimately this meant that, as Stephanson (1995) puts it, 'the puritan break would then eventually serve to invest American nationality with a 'symbology' of exceptionalism or separateness that has survived remarkably intact' (p4).

Given these historical antecedents, it is not surprising that a mythology of the exceptional difference of the space and time of the New World, or what Stephanson (1995) terms 'chosenness,' pervaded the political rhetoric surrounding the birth of the American nation-state. The founders of America independence repeatedly described America as a unique nation, wherein 'America' was the political manifestation of transcendentalism, contrasted over and against the old world (Stephanson, 1995). The Great Seal of America, chosen in 1776, reads, in Latin—'God has blessed this undertaking a new order for the ages' (p5). Similarly, the 1776 Declaration of Independence appealed to the transcendental to animate American exceptionalism—promising 'unalienable Rights, that among them are Life, Liberty, and the pursuit of happiness,' that are 'endowed by their Creator.' Against such transcendental motifs the Declaration of Independence positions the British Crown as the tyrant, whose mode of government stemmed not from the God-given ideals of respect for universal freedom and liberty but from the suggested desire for 'absolute Despotism ... [or] an absolute tyranny over these states.'

Thus, the legitimacy of the formation of the United States, found within the Declaration of Independence, hinges upon a moral distinction between the freedom of imperial governments to rule indiscriminately and the freedom of the individual. In so doing, it explicitly proposes that only the latter society might have an affinity with moral ideality and the divine. On the basis of such moralizing distinctions, the long catalogue of British tyranny, found in the middle section of the Declaration, affirms not simply a sense of political wrongs, but the moral origins of the American people, breaking with the past tyrannies of the Old World and starting out again to work towards moral ideality under God (Anderson, 1991: 193). The Declaration pronounced a nation-state that was founded upon the premise of individual perceptions of material and spiritual liberty being translated into proof of divine affinity with the American nation. This story, commonly referred to as Divine Providence, claimed to identify and instil upon a particular group of people, a path toward messianic salvation; and inserts the Declaration, and the government that followed, as a necessary part of that destiny (Noble, 2002: 4). As a result, the Declaration of Independence's appropriation of Puritanical material and spiritual transcendentalism contributed to a well-established mythology of writing the American continent as not just morally, but spatially and temporally, exceptional in the world. The messianic tenor of the Declaration culminates in the final sentence: 'And with the support of this Declaration, with a firm reliance on the protection of Divine providence, we mutually pledge to each other our Lives, our Fortunes and our sacred Honor.'

With similar transcendental hyperbole, the speeches of early American presidents evoked unspecified transcendental motifs to write a story of exceptional geopolitical difference. Tuan (1993: 204), for example, explains how political and spiritual transcendentalism entered into relation with spatial transcendentalism in precisely this way. Consider a speech given by George Washington in 1788 describing the westward expansion of the US:

> Extent of territory and gradual settlement will enable them to maintain something like a war of posts against the invasion of luxury, dissipation, and corruption. For, after the large cities and old establishments on the borders of the Atlantic shall, in the progress of time, have fallen prey to those invaders, the western states will probably long retain their primeval simplicity in manners and incorruptible love of liberty. May we not reasonably expect that, by those manners and this patriotism, uncommon prosperity will be entailed on the civil institutions of the American world. (quoted in Tuan, 1993: 204).

Tuan (1993) thus observes how 'open space and free land, [became] a source of spiritual and democratic values, including liberty, simplicity and equality' (p204).

While this familiar messianic mythology is palpably evident across eighteenth and nineteenth century American political rhetoric, its political significance must not be over-stated: until at least the twentieth century it functioned, as McDougall (1997) explains, more as an expression of 'what America *was*, at home' (p37, original emphasis) rather than as a rationale for future foreign policy. In other words, exceptionalism operated more or less culturally, binding together a group of people around a purposefully ambiguous destiny, rather than informing specific political strategies. Although the Puritanical mantra of exceptionalism based on freedom from Earthly limits seemed to intuitively herald political unilateralism and imperialistic ambition, during most of the nineteenth century American politicians, such as President Andrew Jackson, were much more routinely concerned with forming strategic alliances to offset European imperial power, developing federal and state bureaucracies to guide economic activity to feed a burgeoning population, or else fighting the Civil War (McDougall, 1997). Indeed, as Stephanson (1995) puts it, drawing on Anderson's (1991) terminology, the mythology of exceptionalism in eighteenth and nineteenth century America should properly be considered as a national 'structure of feeling shared by an *imagined community*' (p28; original emphasis).

Nevertheless, despite the way that this nationalistic appropriation of transcendental destiny appeared absurd alongside everyday political finitudes, complexities and contingencies, it became an increasingly popular self-identity narrative within American popular culture. Here, the ambiguity of a political identity based upon transcendentalism which spoke only of a divinely ordained, though highly ambiguous, destiny for a narrow group of people was easily appropriated as rhetorical justification for many political strategies. Perhaps the most notable instance occurred on 27th December 1845, when John O'Sullivan, founding writer for the popular American magazine *Diplomatic Review*, drew upon a mythology of exceptionalism to justify American expansionism, and in the process coined the now infamous phrase 'manifest destiny' in the *New York Morning News* newspaper. In so doing, O'Sullivan deployed an image of America as exceptional to retrospectively legitimize a divinely ordained destiny to govern the continent from coast to coast under the minimal governmental model typifying Jacksonian democracy. For O'Sullivan, it is American '*manifest destiny* to overspread and possess the whole continent which providence has given us for the development of the great experiment of liberty and federated self-government' (quoted in Stephanson, 1995: 42; emphasis added). His words were in fact a direct response to the imperial ambitions of Britain that had resisted America's annexation of Texas (McDougall, 1997). As evidence to the popularity of O'Sullivan's article, McDougall (1997) explains how for many years within historical and political discourse O'Sullivan became regarded as the 'definitive interpreter of a foreign policy decision' (p77).

An equally popular version of American exceptionalism is found in Frederick Jackson Turner's over-determination of the role of the western frontier in American history, presented at Chicago's World Fair in 1893. The 'Frontier Thesis', as it became commonly known, explains the vast wilderness and 'free land' (Turner does not discuss the territorial claims of Native Americans except as purchases for white farmers wishing to 'civilize' such 'savage' lands) of the Western frontier as territorial evidence of the professedly exceptional narratives of 'democracy', 'progress', and 'freedom', that define American history (Turner, 1893, reproduced in Taylor, 1972: 3-28):

> That coarseness and strength combined with acuteness and inquisitiveness; that practical, inventive turn of mind, quick to find expedients; that masterful grasp of material things, lacking in the artistic but powerful to effect great ends; that restless, nervous energy; that dominant individualism, working for good and for evil, and withal that buoyancy and exuberance which comes with freedom— these are traits of the frontier, or traits called out elsewhere because of the existence of the frontier. Since the days when the fleet of Columbus sailed into the waters of the New World, America has been another name for opportunity, and the people of the United States have taken their tone from the incessant expansion which has not only been open but has even been forced upon them (Turner, 1983 reproduced in Taylor, 1972: 27)

At the core of the frontier thesis is the Puritanical narrative that the American people, and their vast land, possessed an exceptional destiny and identity: spatial transcendence served as an allegory for celestial transcendence. The popularity of the frontier thesis was pervasive both inside and outside academic institutions at the end of the nineteenth century and into the first few decades of the twentieth century (Taylor, 1972).

These nationalistic narratives of the West were complemented by a literary tradition that romanticized frontier landscapes, wilderness and 'Virgin' lands, as the most important expression of the exceptional destiny and identity of the American people. Motifs of exceptional national landscapes pervaded popular nineteenth-century American literature such as: James Fennimore Cooper's *The Pioneers* (1823), Ralph Waldo Emerson's *Nature* (1836) and Henry David Thoreau's *Walden* (1854) (Nash, 1967; or for a geographical perspective: Casey, 2002). In these texts encounters with American wilderness and landscape were frequently read as evidence of affinity between moral ideality, the divine and the American people, precipitating a belief in the exceptional destiny, or what David Noble (2002) calls the 'timeless space' of American nationalism (p1). Roderick Nash's (1967) study of wilderness in American literature concludes similarly that 'by the middle decades of the nineteenth century wilderness was recognized as a cultural and moral resource and a basis for national self-esteem' (p67). Over time, mythologies of American exceptionalism—promises of a transcendental basis for nationalism—found their way through state institutions into popular culture, for example through the American education system. Billig (1995) describes how since the 1880s American school children have been daily required to stand to attention before the national flag and pledge 'allegiance to the flag of the United States of America and to the republic for which it stands, one nation under God, indivisible with liberty and justice for all' (p50). As Billig (1995) argues this ritualized practice of nationalism is often forgotten as such, so that in turn myths of American exceptionalism become cultural potent, yet unremarkable.

Political and popular disseminations of the mythologies of American exceptionalism across American practical and popular geopolitical practices and texts all promise an image of America as the transcendental state. This is not a fully intelligible representation of an absolute sovereign power, with a fixed, territorial border, but rather a somewhat paradoxical outline of a State, where both elements, the transcendental and the State, actively displace and distanciate conventional meanings of the other. On the one hand, this passion for the transcendental draws out the identity and purpose of the American state into the infinite cosmos (in Deleuze and Guattaris' terms a 'deterritorialization'). While on the other hand, concepts of spatio-temporal transcendence become associated with a seemingly fixed and recognizable set of spatio-political practices and institutions ('reterritorialization'), so that the transcendental is no longer regarded as immanent to the omnipotence or omnipresence of God, as in Judeo-Christian theology, or the transcendental reason of the Enlightenment (O'Tuathail's 1996), but rather as immanent to an ethno-politically exclusive ordering of peoples, religions, institutions, materials, ideologies and landscapes.[2] It is with these tensions in mind

---

2   For example, witness Noble's (2002) discussion of how the mythology of American exceptionalism, and hence the image of the 'transcendental state,' naturalized the legitimacy

I deploy the concept of the 'transcendental state'—not as an internally resolved and cohesive representation of an embryonic imperial power, but rather as a *problematic*, that produces and reproduces a rather ambiguous, yet highly potent, technique of writing geopolitical difference, and projecting American geopower.

This unstable image of the transcendental state can be further elaborated by reading it through Derrida's (2002) deconstructive analysis of the Declaration of Independence. Derrida maintains how the enduring, though often overlooked, effect of the Declaration revolves around an undeciability as to whether the Declaration is stating (in Derrida's parlance a 'constative statement') that a people are already 'good' and 'independent' (and we might also add 'exceptional') or whether they are being produced (as a 'performative statement') by the Declaration. Or, as Derrida (2002) explains: 'This obscurity, this undeciability between, let us say, a performative structure and a constative structure is required to produce the sought-after effect' (p49). For Derrida (2002), this undeciability pervades every signature of the Declaration. The 'good' and 'independent' people could only exist once their authorised representatives, who claimed they were simply signing on their behalf, had signed the Declaration, which guaranteed their authority to speak in the name of the 'good people' (Derrida, 2002: 49-50). In this circular manner the signing of the Declaration produced not just the possibility of speaking in the name of the 'good,' 'independent,' and exceptional people, but actually enacts the basis of representative government. What is important, in the context of this study, is the way in which the movement of this endless chain of undeciability was resolved. This was achieved through the countersignature of a final transcendental signing entity, in whose name they appeal to in the final sentence: God. As Derrida (2002) writes here of God, 'He comes, in effect, to guarantee the rectitude of popular intentions, the unity and goodness of the people. He founds natural laws, and thus the whole game that tends to present performative utterances, as constative utterances' (p51). And so, paralleling other deliberations thus far on the transcendental state, it is through faith in the immanence of the absolute transcendental (God) to the State, via notions of the 'good' and 'independent' people, that America as a project of *what ought to be* was legitimated as *what already is*. In this way faith in the transcendental—provided the final recourse through which American state power, national identity, citizenship, morality and representative government could appear united and eternal (Noble, 2002). Yet, as Derrida (2002) suggests, this celestial resolution of the Declaration is unfinished: what happens when we sign in someone else's name; how can we own God's signature?

---

of an Anglo-Saxon elite to govern American political systems.

The American Sublime

While Washington, O'Sullivan, Thoreau and Turner all used words to express a sense of America as the transcendental state, the nineteenth century also witnessed the reproduction, and refinement, of this mythology through paintings of American landscapes. Commonly referred to as the 'American sublime,' this artistic idiom of landscape painting dominated early visual arts movements in the United States throughout most of the nineteenth century. These paintings provided a visualized natural theology, affirming the principle that America was the Earthly vanguard of God's plan, and so humanity's destiny (Barringer and Wilton, 2002, Boime, 1991; Kinsey, 1992; Novak, 1995). This approach to painting American landscapes was practised by numerous artists, including Thomas Cole (1801-48), Edwin Church (1826-1900), Samuel Morse (1791-1872), Asher Durand (1796-1886), Albert Bierstadt (1830-1902), Sanford Gifford (1823-80) and Thomas Moran (1837-1926). The movement can be divided into two stylistically inter-connected groups of artists on the East Coast based around New York (Hudson River School[3]) and the West Coast based around San Francisco (Rocky Mountain School).

The American Sublime was influenced by the European notions of the sublime popularized by Romantic writers and artists, and Enlightenment philosophers, notably Edmund Burke and Immanuel Kant. Thomas Moran (1837-1926), of the Rocky Mountain School, and whom many regard as *the* prominent artist of the American Sublime tradition (Kinsey, 1992), referenced his work to the British aesthete John Ruskin and the renowned British romantic artist J.M.W. Turner. At the heart of the romantic approach to the sublime espoused by Ruskin, and employed by Moran and his contemporaries, lies a way of claiming affinity between the transcendental aggrandizement of landscape, a painter, or viewer's, moral ideals and divine perfection. Ruskin referred to this emotionally elevating[4] resonance as the 'great impression'[5] (Kinsey, 1992; see also Meslay, 2004; Shaw, 2006).

---

3   The 'Hudson River School' was first used as a derogatory term by later impressionist artists at the end of the nineteenth century, in an effort to 'bring back down to Earth' the transcendental aspirations of this early group of painters of landscape (Boime, 1991)

4   The Ruskinan sense of pleasure in the sublime seems to derive from Kant's analysis of the mathematical sublime. While Kant recognized that divine ideals such as moral perfection could not be represented literally, he explained how certain transcendental landscapes in tandem with rational concepts of moral ideals, prefigure a kind of painful failure of imaginative presentation that Kant called negative presentation. In this experience the subject eventually finds pleasure in the sense that these moral ideals can be considered by a rational subject in tandem with nature, even if only partially (Crowther, 1989; Hoffee, 1994).

5   Joni Kinsey (1992) suggests Moran and other landscape artists in nineteenth-century America were drawn to Ruskin's (and in turn J.M.W. Turner's) aesthetic because of his shift towards an aesthetics of nature and morality away from European aesthetic traditions of what was sensuous or acceptable taste, which were 'suspect and unfamiliar in the United States' (Kinsey, 1992: 13). For Ruskin the highest landscape art formed a 'great impression' that pointed to the 'faultless, ceaseless, inconceivable, inexhaustible loveliness,

Moran's emotionally uplifting evocation of this Christianized sublime in the American West was influenced by a number of techniques that were deployed by European painters, such as J.M.W. Turner, as well as the earlier American artists of the Hudson River School. Of particular significance was the use of light effects to convey an impression of natural grandeur and awe. The term 'Luminist' is, in fact, frequently applied to nineteenth-century American, landscape styles (Boime, 1991: 35). Moran's work, for example, often depicts mountains acting as beacons of light or canyons suffused in an incandescent radiance. In so doing, these painters approached the luminous imagery evident in Judeo-Christian mythology, where light is correlated to divinity and hope, and darkness corresponds to wrath and evil.

The relevance of such biblical narratives to American manifest destiny is, perhaps, most explicit in the artwork of Thomas Cole, Asher Durand and Edwin Church, who frequently used light to suggest eschatological narratives of American manifest destiny. Here, the spread of light over darkness accompanied the civilizing passage of Christian pastoral culture into the wilderness (Casey 2002, for example, cites Thomas Cole's *The Oxbow*, 1836, in this regard). And yet, neither Moran nor Bierstadt used light in this way in their depictions of the American West. Indeed, their frontier-scapes had to be set without their pastoral 'Other,' the far-away rural scenes of the Eastern seaboard. Nevertheless, light effects are strikingly pervasive in both the work of Moran and Bierstadt where, as Novak (1995) suggests, light was used rather to augment a process of, '… spiritual transmutation" often by, " … dissolving form,' creating, '… diffusive, vaporous qualities' where, '… light, moves, consumes, agitates and drowns' (p41-2). All of these effects suggest possibilities for spiritual transcendence from the terrestrial to the celestial. A dramatic iteration of the American, Christianized sublime appears when terrestrial forms with precise material detail, such as hilltops, blur into the vast crystalline or transparent light of the radiant sky, evoking the transition from worldly nature to spiritual wonder and awe. Novak (1995: 154) later discerns within these sublime motifs a parallel with early American evangelical readings of the Book of Revelation, where the apocalyptic power of God's creation was to be revealed. It is this revelatory tenor that reproduces the sentiment of American manifest destiny, through the belief that America was chosen by God whose judgment was close at hand, thus augmenting what Novak (1995) terms, '… the American's sense of his own unique nature, his unique opportunity [which] could indeed foster a sense of destiny which, when it served to rationalize questionable acts with elevated thoughts, could have a darker side' (p7).

One of the most compelling examples of the Rocky Mountain School's reading of the American West as a holy text, and by extension an emblem of American

---

which God had stamped upon all things' (Ruskin quoted in Kinsey, 1992: 13). Following J.M.W. Turner and Ruskin, Moran sought to convey an emotional impression of these spaces. He increasingly opposed mere mimetic representation and strove to find a sense of the emotional complexity involved in witnessing these spaces that often gestured towards grand ideals (Kinsey, 1992, Meslay 2004).

manifest destiny, was Moran's *Mountain of the Holy Cross* (1890) (Figure 1.1). At first glance, the painting seemingly imitates the deliberate symbolism of German Romanticism, especially Caspar David Friedrich's *The Cross in the Mountains* (1808). And yet, unlike Friedrich's piece, the cross Moran painted on the mountainside was, in fact, an entirely natural phenomenon. The rock fissures on the apex of this Colorado peak were arranged in a cross that was unmistakable as such when packed with light snowfall. In his depiction, Moran predictably used dramatic light effects to highlight the cross, while also exaggerating the sense of vastness and scale of the mountain, not least by employing a reverential perspective and reducing the height of the woodland in the mid-ground. This evoked a sense of foreboding and trepidation but also the outstanding divinity of the mountain. In this way, Joni Kinsey suggests, both the actual mountain and Moran's rendition became nationalistic icons of the divine redemption of American exceptionalism and destiny with the Godhead, particularly so in the suffering and anguish of the post-Civil War period (Kinsey, 1992: 153). As Samuel Bowles wrote in his book *Switzerland of America* (1869), it is as, '… if God has set His sign, His seal, His promise there—a beacon upon the very centre and height of the continent to all its people and all its generation' (Kinsey, 1992: 149).

The attempt to reconcile the individual, the nation, and God through nature was also practiced through what art historian Albert Boime (1991) terms the 'Magisterial Gaze.' That is, an emotionally uplifting way of seeing the landscape from an elevated perspective. Rocky Mountain School painters, including Moran, frequently employed this technique.

While such a perspective is absent from *The Mountain of the Holy Cross* (1890), which employs a much more reverential upward gaze towards the divine, it is visibly discernable in two of Moran's most famous other works in which the viewer assumes a Godlike gaze over a craggy panorama, *The Grand Canyon of the Yellowstone* (1872) and *The Chasm of the Colorado* (1873-4). Boime explains how this gaze underpins American manifest destiny: '… [the] American experience of the sublime in the landscape occurred on the heights. The characteristic viewpoint of contemporary American landscapes traced a visual trajectory from the uplands to a scenic panorama below … This Olympian bearing metonymically embraced past, present, and future, synchronically plotting the course of empire … [and] remains a fundamental component of the national dream' (Boime, 1991: 1-2). Hence, he suggests, 'It is this systematic projection of the unlimited horizons as a metonymic image of America's futurity that makes this body of material unique in its geographical, national, and temporal setting' (p23-6).

It should be noted, however, that appeals to this God-like perspective as a means of organizing space and time around unifying visions of humanity were not unique to American, romanticist artwork. For example, Dennis Cosgrove (2001) provides an in-depth genealogy of how, '… omniscient and synoptic' (p2) Apollonian perspectives have frequently figured in unifying, eschatological visions. Medieval Atlases of the globe articulate a sense of Christian mission; similarly NASA's Apollo photographs employ Olympian gazes to utopianize

Figure 1.1   Thomas Moran, *Mountain of the Holy Cross*, 45.09 x 31.12 cm. watercolour over graphite on wove paper sheet, 1890,
Source: National Gallery of Art, Washington DC

global internet-age connectivity (Cosgrove, 2001). This Olympian perspective also features prominently in O'Tuathail's (1996) deconstructions of geopolitical discourse-power-knowledge, whereby a disembodied ' ... geopolitical gaze' evokes a Cartesian perspectivism so as to render '... geography spaceless and history timeless.' In consequence, '... both are taken to be transcendental coordinates of the universal nature of things' (Ó Tuathail, 1996: 43) essentializing the histories and destinies of peoples, dramas and places, and, in this case, the American nation. Perhaps what distinguishes the eschatological connotations of the Olympian gaze in the case of American Romanticism is that it appeared against the backdrop of a web of rhetorical connections between divine immanence and religious and national exceptionalism and destiny, from Puritanical divine providence, to Jeffersonian idealism and, later, American manifest destiny (Stephanson, 1995).

Accordingly, it is across these interrelated visual registers of light, composition, symbolism and gaze that the American landscape artwork of the Rocky Mountain School was able to locate the West in the American geographical imagination, through Judeo-Christian mythology and European Romanticism, as the apotheosis of American exceptionalism and manifest destiny. By encouraging the idea that America's frontier geography could be read as an exceptionally holy text, often accompanied by a God-like gaze, these painters reproduced the American West as not just a symbol of the unique destiny of the American nation and its Christian population but as its divine verification. It is against the backdrop of these powerful narratives that these works were viewed and appreciated by the American public and Congress. As Joni Kinsey (1992) puts it, ' ... [through] a world of unspeakable beauty and limitless power ... [they] ... made the West an indelible part of the American consciousness' and in so doing projected 'Christian doctrine onto nature, and by extension onto nationalism' (p176).

The ethno-political heroism of these landscape paintings captivated both the American public and Congress. Indeed, Thomas Moran's earlier oil painting, *Mountain of the Holy Cross*, completed a triptych of three panoramic views of the American West, the first two of which (*The Grand Canyon of the Yellowstone* and *The Chasm of the Colorado*) were purchased by Congress for $10,000 each in 1872 and 1874 respectively. These purchases came after a presentation to Congress of watercolors by Moran and early national parks proponents, including the famed geologist Ferdinand Vandeveer Hayden (Kinsey, 1992). The presentation helped persuade Congress to approve the establishment of Yellowstone as the world's first national park in 1872. Not surprisingly, Moran's sublime landscape vistas were popularized in visitor posters and postcards for early National Parks like Yellowstone (Johns, 2007).

An even more strikingly recognizable and lasting dissemination of the schema of the American sublime tradition was completed in 1945, when Mount Rushmore was carved out of a South Dakota hillside. Boime (1991) describes how the 'great white heads' function as Godhead figures that 'permanently etched into the landscape the magisterial gaze' of the American sublime (p166).

The political appropriation of this visualized expression of national identity endures to this day. For example, both Moran's sold in the 1870s are still owned by the US government and are currently on display at the Smithsonian Museum for American Art (after having been on display in the Capitol Building for over a century). Another Moran, titled, *The Three Teutons* (1895), currently hangs in the White House Oval Office. Back in the 1870s, the early Congressional patronage of Moran's work reflected the high appreciation for Hudson River School art within nineteenth-century popular culture more generally, particularly in the rapidly growing cities of the seaboards (Kinsey, 1992). Indeed, several Hudson River School artists played a central role in either directly founding or providing significant early contributions to most of the major artistic teaching institutions and museums in North America, including New York's National Academy of Design, founded in 1825 by Thomas Cole, Asher Durand and Samuel Morse;[6] the Wadsworth Athenaeum (Hartford, Connecticut), opened in 1844 with major contributions by Thomas Cole and Edwin Church; and New York's Metropolitan Museum of Art, founded in 1870 by Samuel Gifford, Edwin Church and John Kensett (Avery and Harvey, 2003; Howat, 1987; Kornhauser, 2003; TFAOI, 2006). In a similar manner, Thomas Moran and Albert Bierstadt contributed to the burgeoning Californian art scene in San Francisco during the nineteenth century (Hjalmarson, 1999; Jones, 1996). These art institutions continue to display extensive permanent exhibitions of these artists work, in collections that have proved increasingly popular (Peltakian, 2006; TFAOI, 2006); an indication of their popularity is the $35million dollars reportedly paid for a work by Asher Durand in May 2005, which was a then record for an American artist (Rosembaum, 2005).

Even in the nineteenth century these artists sold their pictures for thousands of dollars, as Moran had to Congress in the 1870s. They also enjoyed powerful individual patronage from popular figures, such as the romantic writer James Fennimore Cooper and New York's popular Governor, George de Witt Clinton (Crawford, 1997). These lucrative rewards and influential patrons helped these artists disseminate their work through the founding of art institutions as well as opulent private estates and studio spaces that encouraged the institutionalization of the Hudson River School in America's art community (Crawford, 1997). Though perhaps of more importance than the popularity of this visual culture was the sublime vision of America it institutionalized into the early American art community: a belief in America as the exceptional destiny of humanity, and New York as its pivot. Indeed this is exactly the image of New York City as the 'Empire State' that de Witt Clinton had promised in opening up the riches of the America West by the Erie Canal, turning the city into America's leading commercial centre (Nye, 1994). Hence Nye (1994) proposes that 'the Erie Canal and other internal improvements were components of a "moral machine." They ensured not only prosperity but also

---

6   Samuel Morse was founder and first president of the National Academy of Design, now the National Academy. As his art career failed to develop, in the late 1820's he turned to science and engineering and pioneered the use of telegraphy (Crawford, 1997).

democracy and the moral health of the nation' (p39). The enthusiastic celebrations opening up the Eerie Canal in 1825, and its widespread public support, served to evidence the supremacy of democracy and active citizenship, in short the moral vanguard of American republicanism and democracy (Nye, 1994). Rather than inhibiting sublime sentiments of Nature, Nye (1994) explains how these early technological projects engendered a distinctly American sublime: 'Not limited to nature, the American sublime embraced technology. Where Kant had reasoned that the awe inspired by a sublime object made men aware of their moral worth, the American sublime transformed the individual's experience of immensity and awe into a belief in national greatness' (p43).

Within critical geopolitical approaches, the universalization of articulations of global space, organized through geographical imaginations, is understood to be synonymous with the practice of formal and popular geopolitics. As O'Tuathail (1996) put it, 'Geopolitics … is precisely about moral claims and deep interpretations that postulate a fixed and homogenous essence, an underlying law, a relentless continuity to international politics' (176). American manifest destiny, as envisioned through the Rocky Mountain School, provided exactly that underlying law and continuity to the geography of global space. In its broadest sense, this process was enacted through a circular logic: artists journeyed into the American West carrying with them Puritanical narratives of the exceptional destiny of America in the world, which they subsequently confirmed through a particular way of seeing and representing the landscapes of the American West. Thus these artists provided a definable, recognizable and repeatable aesthetic framework that translated the 'frontier' landscapes of the American West into a stage upon which an exceptional, and exclusive (white, Christian), version of American identity could be performed. However, as McDougall (1997) makes clear, during the nineteenth century, a great deal of this messianic mythology was just that—it had a minimal bearing on foreign or domestic policy—but as the twentieth century progressed, the separation of American myth and politics dilutes. And it is one less often discussed register for the spread of this messianic self-identity into practical policy-making that will constitute the focus of my analysis in the next chapter.

Chapter 2
Framing a World Beyond[1]

Cosmic Dreams

In the preceding chapter I introduced how motifs of transcendence have long held sway over those seeking to explain what America *is* and *will become*. In this chapter I examine how this seductive, but quixotic, form, of inscribing exceptional America's place in the world, precipitated a particular way of seeing, feeling, thinking, and eventually experiencing, outer space. Central to this process were the diverse ways through which motifs of the transcendental state percolated through popular space cultures during the 1950s; of particular importance here is the space artist, Chesley Bonestell, and his relationship with various corporations and individual space proponents. After the launch of Sputnik, and early Soviet space successes, popular space cultures offered a nationalist pretext for President John F. Kennedy to pursue an ambitious program of human space exploration as part of a familiar story of American identity and destiny. In turn, these transcendental passions were assembled to organize what became an immense American drive for space exploration. These themes are the focus of this chapter.

The visualization of the transcendental state in outer space within American popular imaginations was mediated largely through one man's fascination and talent for painting outer space: Chesley Bonestell (1888-1986). Bonestell's rise to eminence began in 1952, when the architect turned film effects designer and part-time American space artist began working with the acclaimed rocket scientist Werner von Braun to produce illustrations for a *Collier's* magazine series titled the 'Man Will Conquer Space Soon!.' Von Braun and Bonestell's work reflected a shared passion for the immanent prospects of human space travel to other worlds and, in Bonestell's case, a particular interest in conveying this fascination to others through the image. The two men themselves recognized how their interests and skills complemented each other: for example Bonestell is quoted in admiration of von Braun's, ' ... intellect, romanticism and modesty' (Miller and Durant, 2001: 73), while von Braun spoke of how he, ' ... learned to respect, nay fear, this wonderful artist's obsession with perfection' (Miller, 2007).

Congress and President Eisenhower, however, appeared much more interested in von Braun's rocket technology for its strategic value in delivering a nuclear warhead or a surveillance satellite (Burrows, 1998). Nevertheless, together von

---

1   Material in this chapter was first published in 'Framing Space: A Popular Geopolitics of American Manifest Destiny in Outer Space,' Sage, D, (2008), *Geopolitics* (Routledge), reprinted by permission of the publisher (Taylor & Francis Ltd, http://www.tandf.co.uk/journals).

Braun and Bonestell looked to augment popular support for space travel to help persuade the American government of the merits of a more ambitious, human space travel program. The space historian Howard McCurdy (1997) explains how Bonestell's images played a vital role in this effort to promote space travel:

> No artist had more impact on the emerging popular culture of space in America than Chesley Bonestell ... Bonestell did for space what Albert Bierstadt and Thomas Moran accomplished for the American western frontier ... Bonestell's paintings took viewers to places they had never been before ... [and so] created a sense of awe and wonder ... he used light and shadow, as artists had done with the American west a century earlier, to portray space as a place of great spiritual beauty. Through his visual images, he stimulated the interest of a generation of Americans and showed how space travel would be accomplished (p45).

Space art commentators (for example Hine, 1992; Hardy, 2002; McIntyre, 1986) have observed how Bonestell's lunar surfaces manifestly resemble the intricate craggy rock formations of, for example, the Rocky Mountains that are so indicative of the American sublime tradition. The question that arises is: how did Bonestell's images work to imaginatively locate outer space, and in particular the Moon (as will be examined in Chapter 3), under the same nationalistic and moralizing rubric as the 'frontier-scapes' of the American West? And how did these images popularize travel to these otherworldly landscapes as the next, logical chapter in American exceptionalism, futurity and destiny?

While other artists, such as Charles Bittinger in *National Geographic* (1939) and Lucian Rudaux in the book *On Other Worlds* (1937) (McCurdy, 1997; Miller and Durant, 2001), had attempted earlier, realistic depictions of outer space and space travel to accompany non-fiction works, the widespread popularity of Bonestell's images mark out the significance of his work. Specifically, it was the *Collier's* series that was to prove most effective in disseminating Bonestell's particular way of seeing outer space, not least because *Collier's* magazine was placed in the top ten of American magazines with a weekly readership of over three million people (McCurdy, 1997). As space art reviewers Ron Miller and Fred Durrand (2001) suggest, 'The *Collier's* articles were the beginning of the Golden Age of spaceflight—that period during which the American public showed a fascination, enthusiasm and support for spaceflight it had never shown before or since—the astronautical equivalent of the aviation craze of the 1920s and 1930s' (p77). Bonestell's work in *Collier's* overshadowed that of the two other artists engaged, Rolf Kelp and Fred Freeman, becoming irreducibly associated with *Collier's* eight-part story of space exploration, from the construction of a space station to a Moon base and eventually a Mars expedition (McCurdy, 1997: 40). While Kelp and Freeman prepared space-station cutaways and rocket ships, Bonestell depicted vast lunar vistas flanked by craggy mountains, with Moon bases or space stations with massive centrifuges floating high above the Earth (Figure 2.1).

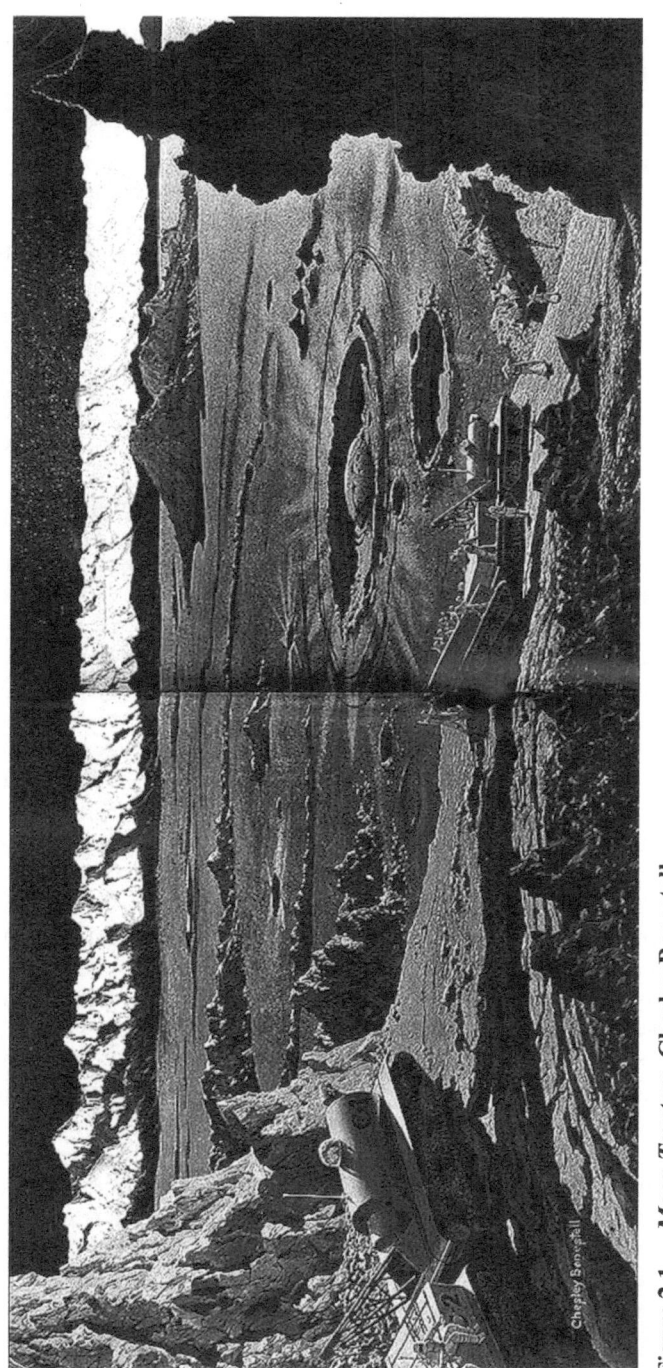

Figure 2.1 *Moon Tractors*, Chesley Bonestell.
Source: Collier's magazine, October 25th 1952 © Reproduced with kind permission from Bonestell Space Art

The *Collier's* series was complemented by three popular, non-fiction books illustrated by Bonestell, *Across the Space Frontier* (1952), *Conquest of the Moon* (1953) and *Exploration of Mars* (1956), as well as several commissions for large murals by major US art galleries and museums. In the aftermath of the *Collier's* series, the Walt Disney Company recruited Willy Ley and von Braun, amongst other space boosters, to create a series of animated television shows that told the story of outer space.

The shows were partly intended to attract popular attention around the 'Tomorrowland' element of the soon to be opened icon of American popular mythology: the Disneyland theme park in Anaheim, California (which opened in 1955). Bonestell's images were used as the basis for many of the images, animations and cartoons that were employed in Disney television series and, indeed, were to influence design themes in the Tomorrowland exhibit itself (McCurdy, 1997: 41-3). The *TWA Moonliner*, for example, was an attraction demonstrating how by 1986 space travel would be as commonplace as air travel in the 1950s. The attraction strikingly resembled a V2 rocket and was designed by von Braun with financial support from Howard Hughes of TWA airlines. In consequence, as acclaimed space writer Arthur C. Clarke explains, Bonestell 'electrified a generation of teenage space enthusiasts, aspiring writers, physicists, artists and engineers' (quoted in Miller and Durant, 2001: 9).

Bonestell's vision of outer space is all the more distinctive because of the contrast it struck with contemporaneous images of outer space. During the 1950s, Hollywood sci-fi regularly constructed dystopian images of outer space as another setting for the movie horror genre; it was a fearful realm full of warmongering aliens (Booker, 2001 and Hendershot, 1999). Despite Bonestell's work on films such as *War of the Worlds*, his ostensibly non-fictional representations of outer space provided an alternative, wherein outer space could be habituated as the utopian or heroic destiny of America and humanity. As a result, the geopolitical connotations of Bonestell's romantically uplifting vision of space exploration resists equivalent interpretation as a metaphor for Cold War anxieties, such as nuclear Armageddon, anti-communism or the de-humanizing effects of technoscience. Instead of evoking these anxieties directly, Bonestell's detailed images presented a much more comforting and structured view of the universe and the progressive role of technoscience, which may well have helped to allay such Cold War anxieties. The rocket, for example, was readily familiarized as an emblem of national futurity and progress instead of war and destruction (cf. Macdonald, 2008). Augmenting this sense of order was a frame of authority: Bonestell's vision was presented as a factually determinate, even quasi-scientific, estimate of what these other worlds would look like and how space travel would take place. For example, as Willy Ley explains of Bonestell's art in the *Conquest of Space* book of 1950, ' ... [they] should not be considered "artistic conceptions" in the customary sense of the phrase ... but a picture which you might obtain if it were possible to get a very good camera with perfect color—true film into the proper position' (Ley, 1950: 9-10). And yet it is precisely this frame of objectivity that makes Bonestell's work all the more geopolitically significant, as such neutrality and detachment belied its innate equivalence to the ideologically and morally charged approaches of the Rocky Mountain School and American Romanticism.

These similarities are rendered all the more remarkable because Bonestell himself was a self-declared atheist who no doubt would have rejected the thought that his work was acquiescing, even if inadvertently, with the explicitly evangelical idiom of American landscape art. And yet, as commentators on space art acknowledge (McCurdy, 1997 and Miller and Durant, 2001), the implicit inter-textuality between Bonestell's astronomical art and the geographical imaginations reproduced in artwork from the Rocky Mountain school is central to disentangling its socio-cultural context, popular reception, and, in turn, its geopolitical significance. Two interrelated visual incongruities in Bonestell's representation of the Moon (Figure 2.1) help us to approach the more implicit, popular geopolitical connotations present in his work. The first relates to the lunar landscape and the 'problem' of depth. After walking on the Moon the Apollo astronauts repeatedly reported problems judging distance, depth and scale on the featureless and bland lunar landscape (for empirical accounts from Apollo astronauts see Chaikin, 1998). By contrast, Bonestell employed an Olympian perspective that conveyed an immediate sense of scale, vastness and immensity. However, given Bonestell's contact with leading space scientists and interest in astronomy he surely would have known that judging scale on the relatively featureless lunar surface would have been much more difficult. When asked about his use of perspective, Bonestell explained how he 'used the device of near rocks and distant mountains, separated by a plain glimpsed beyond the foreground, to give the impression of depth and distance' (quoted in Miller and Durant, 2001: 109-10). More inconspicuously, and echoing Bierstadt, Bonestell often placed tiny human figures in the foreground of his paintings to heighten a sense of scale and immensity. The importance of this enhanced sense of depth from an elevated perspective is that it inherently approaches many of the techniques found in the Rocky Mountain School (and American Romanticism generally). To re-iterate, this Olympian perspective had been long employed to correlate natural feelings of terrestrial transcendence, uplift and wonder within moral and religious frameworks (Cosgrove, 2001). Moreover, as Boime (1991) acknowledges, this Olympian perspective also had important geopolitical connotations in the context of American culture, providing a sense of the artist as occupying a 'heightened position,' surveying the past, present and future horizons of humanity itself.

The second visual incongruity concerns the aesthetic implications of Bonestell's fondness for an innately 'Olympian gaze.' That is, Bonestell required a strong foreground and background in his paintings in order to provide a sense of perspective that otherwise would have been sorely lacking if he had indeed depicted a rather flat and smooth surface of the kind that Apollo astronauts, in fact, reported. As Bonestell himself discloses, his decision to depict lunar mountains as craggy and sharp was a necessary bi-product of his use of an Olympian perspective. In a 1977 interview Bonestell attempted to explain why he took the decision to paint the lunar landscape the way he did, despite the fact he almost certainly knew it was much smoother: 'I don't think it [the Moon] would have looked very interesting, although it would have been correct. I tried to make it just as dramatic as I could' (quoted in Miller and Durant, 2001: 28). Bonestell later said on the same subject, ' ... even if they're wrong, they did influence young people and got them interested in astronomy, so they at least served that purpose' (quoted in Miller and Durant, 2001: 93). Here, Bonestell reveals that underneath his ostensive realism he was acutely aware of how different landscapes might evoke different sentiments and indeed emotions in his audience. By painting in this romanticist style he sought to engage an audience's familiar expectations of what a seductive and awe-inspiring frontier landscape should look like. Bonestell was drawn to craggy, alpine-esque scenery framing vast planes as the most romantic, alluring, uplifting and seductive landscape to inspire space travel. And yet, it must be noted, Bonestell himself never formally acknowledged the influence of the Rocky Mountain School, despite the fact that many of these earlier artists were hugely influential in the founding of the San Francisco art community where Bonestell grew up and developed his technique (Hjalmarson, 1999). Conceivably, this decision might reflect his unease with this earlier, overtly religious tradition of landscape art.

Beyond these aesthetic techniques, the text in *Collier's* reinforces the popular, geopolitical connotations implicit in Bonestell's images, as Americans were told that, 'Man will conquer space soon,' first creating a space station, then a Moon colony and, in time, undertaking a Mars expedition (McCurdy, 1997: 40). The contours of this popular roadmap, as illustrated by Bonestell, with its focus upon human colonization, predominated over and above alternative 'non-human' activities in outer space, such as the use of robotic scientific probes and military satellites. The innate similarities to nationalistic mythologies of manifest destiny in the American West, where frontier exploration went hand in hand with physical settlement, were strikingly visible in these popular visions of space travel. Not surprisingly, countless replicas of Bonestell's frontier vision of space exploration appeared across the world during the pre-Apollo years, including within innumerable books for children. As Miller and Durant (2001) illustrate: 'The Collier's spacecraft, and even the artwork itself, were copied and plagiarized endlessly. If anyone had to illustrate a rocketship it had to look like a Collier's rocket or it just wasn't right. They were the standard' (p77).

Bonestell's images of outer space implicitly coordinated outer space as the 'high' or 'new' frontier in a purportedly unified American imagination so as to help familiarize an otherwise ominous environment around a pre-established nationalistic mythology. Jonathan Smith (2002) explains how such frontier landscapes are integral to the articulation of American identity, since 'It is of course, impossible to pretend that the American people sprang from common ancestors, from a mythic tribe in the midst of antiquity, as so many other nations do, and so it is necessary to define the group by its relation to a common territory [the frontier]' (p321). By familiarizing outer space in these eschatological and exceptionalist terms, Bonestell naturalized the American 'frontier' paradigm, of human space colonialism, as not just the most likely course of the space age but as part of the performance of American national identity in the Cold War. Such a mobilization of the popular cosmographical imagination dovetails with Sharp's (2000) insistence on how popular geographical imaginations are deployed by interest groups to augment national mythologies of belonging that familiarize new and threatening situations and naturalize certain political assumptions through ' … accepted models, metaphors and images' (p335); thereby instilling a sense of pre-destiny to foreign policy decisions and outcomes.

New Moon Rising

Bonestell and von Braun's vision of an American, and indeed universal, future being staged through space exploration of outer space reached a turning point in October 1957—just over three months into the first International Geophysical Year, a year which for many promised the launch of the first orbital satellite by America—yet, instead on the 4th October the USSR successfully sent a 184 pound satellite into Low Earth Orbit. The satellite called 'Sputnik' ('traveling companion' of the Earth), was reported across the American news media to suggest a tragic defeat, comparable to Pearl Harbor, not just of American technological prowess but also political ideology and American society (Burrows, 1998: Dickson, 2001; Launius, 2005; McCurdy, 1997). Two days after the launch, the *New York Times* described how 'In the Soviet mind the launching of the earth satellite was not a victory of man over nature but of "Socialist society"' (Jorden, 1957). The reaction of Congress was equally intense. Senator Henry. M Jackson, for example, was reported in the *New York Times* suggesting that Sputnik undermined American capitalism, dealing 'a devastating blow to the United States scientific, industrial and technical prestige in the world' (Mooney, 1957). A similar sense of defeat pervades a *New York Herald Tribune* editorial (5th October, 1957) that concluded: [this is a], 'grave defeat for America' (Anon, 1957a). In fact many of these anxious responses were precipitated, in part at least, by the original Soviet news communiqué from Soviet news agency, *TASS,* whose report of the Sputnik praised the event because: 'Artificial earth satellites will pave the way for space travel and its seems that the present generation will witness how the freed and conscious labor of the people of the new Socialist society turns even the most daring of man's dreams into a reality' (full transcript available in Logsdon et al., 1995: 329-30).

Some reports were noticeably more upbeat and tried to resolve the crisis by re-stating a von Braunian-Bonestellian vision for an American-led conquest of space. An article in the *New York Times* laid out a vision for future space exploration seemingly lifted straight out of von Braun's plans in *Collier's,* with the corresponding title: 'New Space Conquest Can Now Be Foreseen' (Plumb,

1957).[2] And yet in spite of these moments of optimism, the overall tenor of the media reaction was of some kind of defeat of American universal destiny and exceptionalism. Set against the way the image of America as the transcendental state had been projected into outer space by von Braun-Bonestell this was perhaps hardly surprising. Working within the 'logic' of this messianic mythology was a potent conflation between spatial and temporal frontiers, and so the media compared Sputnik 'to the discovery of America by Columbus' (McCurdy, 1997: 62-3). Here the claiming of a new spatial frontier was readily understood as a cipher for the claiming of humanity's universal destiny.

The American press was also at pains to report how the launch of Sputnik challenged perceptions across the world of American claims to be leading humanity's moral and temporal, as well as spatial, vanguard (Dickson, 2001). The *New York Times* reported on October 7th 1957 that both Chinese and Austrian communist newspaper's saw Sputnik as an 'epoch-making achievement proving the superiority of Soviet science,' while Vienna's newspaper, *Communist Volkstimme*, quoted the German writer, Goethe, and declared, 'Here and now begins a new epoch in world history' (Anon, 1957b)—once more a spatial frontier was also being put forward as a temporal frontier. A *Washington Post* article assessed that the effect of Sputnik

> ... will be greatly to enhance Soviet prestige and to buttress the Moscow claim that communism is the wave of the future. Americans would only deceive themselves by drawing any conclusions other than that the Soviet Union has scored a brilliant victory in the race for scientific progress and for leadership in the ideological sphere' (Anon, 1957c).[3]

Simultaneously, Sputnik also tapped into the growing anxiety that American freedom from external threats might no longer be assumed. Moran's 'magisterial gaze' (Boime, 1991)—where America's soil was an expression of self-government and control under the watchful eye of an American Godhead—appeared to be withdrawing from the Cold War scene; it was also reported as inevitable, though

---

2   This said, despite the von Braunian rhetoric, the media tended to focus upon the official American civilian satellite program: *Vanguard,* an untried craft constructed by the Naval Research Laboratory, and scarcely mentioned von Braun's army rocket team and their more flight-proven Redstone or Jupiter C rocket, largely because in 1955, Eisenhower had chosen von Braun's team for military ICBM development rather than as a possible first step in American space exploration (see Dickson, 2001: 87).

3   Sputnik documents connections between the image of America as the transcendental state and popular beliefs in the kind of Self-Other binary that circulated during the cold war as a struggle between competing universalizing moralities and ideologies, not just within popular geopolitics (for example Sharp, 2000); but also practical geopolitics. The *National Security Council* meeting of 1968, spoke of the Soviet Union expounding 'a new fanatic faith antithetical to our own,' to seek 'absolute authority over the rest of the world' (Stephenson 1998: 80).

in fact incorrectly (Burrows, 1998: 187), that Soviet ICBMs might now reach American soil unimpeded (McCurdy, 1997). Similarly, an article published in the *Washington Post* (October 13th, 1957) examined the implications if Sputnik has eyes, reporting 'For if Sputnik has eyes, Sputnik provides the means to zero in the Soviet instruments of total destruction on every important American target' (Alsop, 1957).

The emphasis on the visual menace of Sputnik throws into relief the way the Olympian visual motifs of the transcendental state were being directed towards outer space. The historian, Susan Cottler, recalls her experience of such fears: 'Everyday people—teenagers, people in their 20s, parents—would look up in the sky at night and try to spot it. They would wonder if they had to speak in hushed tones, or have arguments behind closed blinds, for fear that Big Brother was, in fact, spying on us' (Cottler, quoted in Dickson, 2001: 113). Further incarnations of this Olympian omnipresence appeared in cartoons. For example, the cartoon 'Moonglow' (Figure 2.2) published in the *Washington Post* on October 15th 1957, ridicules President Eisenhower for cowering in his bed as the light of the 'second Moon,' as Sputnik was often called, appears to herald the dawn of a new age of Soviet supremacy.

What is particularly interesting in this cartoon is the use of the high ground and light—both motifs accompany the articulation of the transcendental state in nineteenth-century landscape art (Boime, 1991). A similar sense of fear gripped Senator Richard Russell, chairman of the Armed Services Committee, who was reported as saying—'From a military standpoint, it confronts us with a new and terrifying danger ... [and so] ... we must not become so fat and lazy that we lose our system of self-government through neglect' (quoted in Anon, 1957d). Such fear, as Dickson (2001) surmises, was inherently spatial and visual: 'There it was overhead-visible to the naked eye and audible to anyone with a shortwave receiver. America had fought two world wars, protected by the breadth of oceans and the comfort of a strong Navy. A certain sense of invulnerability seemed to be an American birthright' (p128).

In the wake of the shock of Sputnik, President Eisenhower's administration attempted to dispel the growing public anxiety—over both the loss of prestige and the perceived threat of Sputnik's visual omnipresence, and threats of nuclear attack—with hope of future American steps forward in space technology, combined with caution over the significance of Soviet advances (McCurdy, 1997). Eisenhower unequivocally declared 'Our satellite program has never been conducted as a race with other nations' and that it did not increase national security threats 'by one iota' (quoted in Lawrence 1957). And yet, given the historical significance of the transcendental state as a motif of American geopower, and then the work of Bonestell and von Braun in harnessing outer space to express these mythologies, a degree of public paranoia was perhaps understandable. Indeed, in an annex to a now declassified National Security Council policy report, which President Eisenhower endorsed in 1955, there is repeated reference to the psychological or propaganda value of the first launch of a satellite, exemplified in

Moonglow

Figure 2.2 *Moonglow*, **Herblock**
Source: Washington Post, 15th, October 1957 A 1957 Herb Block Cartoon © The Herb Block Foundation

the passage: 'The stake of prestige that is involved makes this a race we cannot afford to lose' (available in Logsdon *et al*, 1995: 313).

In this manner, Sputnik became a 184 pound extraterrestrial catalyst for paranoiac affect, as it looked down upon a nation organized around faith in its universal transcendence. Sputnik provided perhaps a fittingly paranoid emblem for the Cold War, a war, which Anders Stephanson (1995) describes, was fought over 'ideological, political, and economic claim[s] to universality' (p83). Dickson (2001), concludes similarly: 'The satellite touched off a superpower competition that may well have acted as a surrogate contest for universal power – perhaps even a stand-in for nuclear world war' (p6). Through the image of America as the transcendental state, Sputnik was read through a range of what William Connolly (2002), terms 'somatic markers.' These are, as Connolly (2002) explains 'preliminary [i.e. non-cognitive] orientations to perception and judgment' which 'scale down the material factored into cost-benefit analyses, principled judgments, and reflective experiments' (p35). In the case of Sputnik, what appeared to be result of these affect-imbued[4] somatic markers, in many cases, was a kind of downscaling of the open-ended potential of Sputnik into polarized (and so mutually dependent) emotionally charged judgments of messianic hope versus eternal damnation. These somatic markers, underpinned by the extension of the transcendental state into outer space in 1950s popular culture, served to geopolitically downscale Sputnik into an apocalyptic conflict, so that, in turn, it could become an analog for a war between moral universals (compare with Dijink, 2006). Tom Wolfe perhaps best expressed the simplifying Sputnik affect in his popular novel, *The Right Stuff*, writing, 'Nothing less than control of the heavens was at stake, [Sputnik] was Armageddon, the final and decisive battle between good and evil' (Wolfe, 1979: 16).

Sputnik appeared to threaten two registers of the image of America as the transcendental state: visual and technological. First, Sputnik looked down and traversed the ground, like the American Godhead of the magisterial gaze (Boime, 1991). The media picked up on this panoptic aura, as in the cartoon of Eisenhower cowering in bed (Figure 2.2). Second, Sputnik's technology transgressed the material limits imposed by Earth's gravity; a threshold which American scientists and engineers had not overcome. In turn, Sputnik directly challenged the view of the American technological prowess passing over the horizons of 'Nature' to claim humanity's future (Nye, 1994), exemplified in nineteenth-century artistic works such as John Gast's *American Progress*:

Even the renowned American physicist Lloyd V. Berkner hinted at this new age of human control of nature, by suggesting that (Figure 2.3 below).

> From the vantage point of 2100 A.D., the year 1957 will most certainly stand in history as the year of man's progression from a two-dimensional to a three-dimensional geography … The earth satellite is a magnificent expression of man's intellectual—growth of his ability to manipulate to his own purposes the very laws that govern his universe (quoted in Dickson, 2001: preface).

---

4 The term 'affect' is used here in the way Connolly (2002) employs it, drawing on Damasio, as the half-second that occurs to prime thought to certain patterns of judgement and cognitive behaviour. In Chapter 8 the concept of 'affect' is elaborated in more detail through the work of Deleuze and Guattari.

Sputnik was translated into a portent for an apocalyptic battle for the control of cosmic space and time. This in turn set the scene for increased public and Congressional eagerness for exactly the grand image of the conquest of space as outlined in von Braun and Bonestell's *Collier's* series.

In response to the feeling of menace Sputnik provoked within American popular geopolitical imaginations, Congress and President Eisenhower unsurprisingly decided upon the establishment of a more pro-active space program. Accordingly, Werner von Braun's team was chosen to quickly launch a small satellite that they had first proposed (with negative replies from Congress) in preparations for the IGY (McCurdy, 1997). Eventually, on January 31st 1958, von Braun and AMBA launched America's first satellite, Explorer 1 (Burrows, 1998; Launius 2005). Tellingly for von Braun it was this launch, not Sputnik or the even more advanced Sputnik 2 that was 'the beginning in the long-range program to conquer outer space' (von Braun quoted in Dickson, 2001: 177). The messianic tone of this re-vitalized American effort was evocatively captured by John D. Medaris, a military leader of the AMBA: 'The overriding importance of space exploration takes on proper significance only if we appreciate that it will make possible a new understanding of man's relationship with the infinity of Divine Creation' (quoted in Dickson, 2001: 178). In spite of the professed technical and spiritual success

**Figure 2.3   John Gast, *American Progress*, 32.5 x 40.6cm, oil on
            canvas, 1872**
Source: National Library of Congress, Washington DC

of Explorer I, and in due course other spacecraft such as Vanguard, there was a growing consensus of opinion in Congress and from President Eisenhower that the US required greater direction and effectiveness in civilian space activities, not least autonomy from military bureaucracy (Dickson, 2001). Eisenhower hinted at the direction he foresaw outer space must fulfil in a report produced by the President's Science Advisory Committee entitled *Introduction to Outer Space* (Logsdon *et al*, 1995: 332), released on March 26th, 1958. In this report the image of America as the transcendental state is evoked subtly in a section which asks the question—'What is the purpose of a national space program?' The first answer given is the 'compelling urge of man to explore and to discover; the thrust of curiosity that leads men to try to go where no one has gone before' and then, 'there is the factor of national prestige. To be strong and bold in space technology will enhance the prestige of the United States among people of the world and create added confidence in our scientific, technological, industrial, and military strength' (Logsdon *et al*, 1995: 332). Between these two answers we are led to the idea that America's national identity is already bound up with the cosmos: human exploration of the unknown.

These policy pronouncements eventually culminated in the establishment of the National Aeronautics and Space Administration (NASA), with the former National Advisory Committee for Aeronautics (NACA) at its core.[5] NASA was established in law by the National Aeronautics and Space Act of 1958, signed by President Eisenhower on July 29th, 1958; NASA itself became an operational governmental agency on October 1st 1958 (Burrows, 1998: 218). The 1958 space act is a rather anodyne document that was much less explicit about the relationship between nationalism and space exploration than earlier proclamations by Eisenhower, Congress or the American media. Across eight policy declarations, the most palpable example of the conflation between American space exploration and humanity's future appears in the passage: 'The Congress hereby declares that it is the policy of the United States that activities in space should be devoted to peaceful purposes for the benefit of all mankind' (NASA Space Act 1958).

In contrast, the now declassified, space policy document, *US Policy on Outer Space* (1960) evidences a far less equivocal articulation of America as *the* transcendental state. In this policy document, reference is made to the 'psychological impact of outer space activities which is of broad significance to national prestige' and the need to exploit American psychological advantages by funding space projects that offer a clear American lead. Moreover, the report warns that Soviet 'firsts' (by January 1960, the USSR had launched three Sputnik spacecraft—Sputnik 2 alone was significantly heavier than all seven US Explorer

---

    5   The NACA was established in 1919, to conduct research into aeronautics both for industry and the military. Its success at this task was mixed, and it was deemed too narrowly focused to develop space activities while remaining the NACA, and so the National Aeronautics and Space Act of 1958 (Logsdon *et al*, 1995b: 334-45) replaced the NACA with NASA (Burrows, 1998: 217-8).

and Vanguard spacecraft combined, and contained the first living passenger—a dog, 'Laika') have confirmed Soviet 'credibility' so that the Soviet leaders can claim, 'general superiority ... [and a] world balance shifted in favour of communism' and most importantly, in light of this discussion, the USSR can assert, 'Communism is the wave of the future' and that it is a, 'sophisticated nation that is equal to the US in most respects, superior in others, and with a far more brilliant future' (available in Logsdon *et al*, 1995: 362-73). Here, in a US space policy document, we find a clear account of how space exploration was being understood by American politicians as a benchmark to assess the ideological claims of rival superpowers.

Kennedy's Vision

While it was the Eisenhower administration that established NASA, under Eisenhower's leadership NASA's aims were cautious and divergent, serving the pragmatic interests of commerce, science or the military, as opposed to the kind of ambitious space-exploration program suggested by von Braun or Tsiolkovsky. Despite various space policy documents warning Eisenhower that public opinion was increasingly permeated by beliefs that outer space was vital to America's identity, the President was keen not to unbalance the budget and undermine other government policy priorities (Burrows, 1998, Dickson, 2001, Logsdon *et al*, 1995). Rather, it was the Democratic Party, and in particular Senators John F. Kennedy and Lyndon B. Johnson, who sought to engage the American public through an image of American space exploration as the destiny of the transcendental state. An exemplary performance, in this regard, was the Democratic presidential election of 1960, where Kennedy and Johnson played upon the popular post-Sputnik belief that America required a more ambitious space program because this was a matter essential to national security or indeed national survival and that, importantly, only the Democratic party was prepared to win this 'space war' and fund such a project (McCurdy, 1997). Kennedy and Johnson both promoted the idea that command of the space and time of humanity would be determined through the possession of the most advanced program of space flight and space technology (for more on this see Burrows, 1998: 189, 320-1, Beschloss, 1997: 51-3; McCurdy, 1997: 75-6; McDougall, 1985: 389-402). During his 1960 Presidential Campaign Senator Kennedy, for example, exclaimed to the electorate—'Control of space will be decided in the next decade. If the Soviets control space they can control earth, as in the past centuries the nation that controlled the seas dominated the continents ... .We cannot run second in this vital race. To insure peace and freedom, we must be first' (quoted in, McCurdy, 1997: 75).

Kennedy's vision was well received. It seemed to offer a new stage for America to recognize its transcendental mission. And, in so doing, also laid out a new mantle for the magisterial gaze looking down on the Earth from outer space, from the vantage point of God (Boime, 1991; Cosgrove, 2001). Such speeches proved highly popular, and no doubt helped Senator Kennedy win the Presidential election of 1960 against incumbent Vice-President, Richard Nixon (Burrows, 1998: 319). Staying true to his campaign speeches—and in the wake of the successful orbit of the first human in space, Yuri Gagarin, by the USSR on April 12th 1961—on May 25th, 1961, in his second State of the Union address to both houses of Congress, Kennedy presented a radically ambitious vision of American space exploration, which centered upon a plan for American human lunar exploration by the end of the decade. While the speech built upon earlier comments made by Kennedy and Johnson regarding the necessity for America to take a lead in outer space, it more directly emerged out of a consultation review. This review, instigated at President Kennedy's request by Vice-President Lyndon Johnson, asked various individuals inside and outside government, for their views on whether and whether 'we have a chance of beating the Soviets?' (Kennedy, 1961a). The result was a range of responses that themselves seem to have been lifted straight out of von Braun's ambitious manifesto for space exploration, as outlined in *Collier's* magazine in the 1950's (Beschloss, 1997: 56-7). The centerpiece of the plan was a proposed, manned, lunar landing through a plan that echoed the comments made by von Braun post-Sputnik, that America must and can take a lead in outer space. The *Collier's* feel of Kennedy's vision was hardly surprising as by this point Werner von Braun was employed as Director of NASAs Marshall Spaceflight Centre and his ideas were included within Johnson's review (Beschloss, 1997: 57). In his speech, Kennedy framed the Cold War as a universal battle between 'freedom' and 'tyranny'—activities in outer space were influential in deciding the winner of this battle. He argued:

> the impact of this adventure [into outer space] on the minds of men everywhere who are attempting to make a determination of which road they should take … .Now it is time to take longer strides—time for a great new American enterprise—time for this nation to take a clearly leading role in space achievement which, in many ways may hold the key to our future on earth.

> This nation should commit itself to achieving the goal, before this decade is out, of landing a man on the Moon and returning him safely to earth. No single project in this period will be more impressive to mankind or more important for the long-range exploration of space. (Kennedy 1961b).

Here Kennedy framed his words carefully to write America as the leader not just in space technology but to some extent the destiny of humankind, further conflating American achievement with humanity's universal destiny. Support for the plan was nearly unanimous in a vote in Congress; in no small part this was due to Johnson's earlier efforts in building a cross-party consensus (Beschloss, 1997: 61).

Numerous space commentators have tried to identify why the decision by Kennedy was taken to support the lunar-landing project, subsequently the Project Apollo (Beschloss, 1997; Burrows, 1998; Dickson, 2001; Logsdon, 1969, 1994, McCurdy, 1997). This is an important question for the present study too, because the American decision to opt for a lunar landing sheds light on how the problematic of the transcendental state entered into American space policy. Dickson (2001: 216-7) outlines four competing rationale for Project Apollo:

1. Imperialism—the idea that control of Moon equals control of the space and time of the Earth (as in Kennedy's speech to congress in 1961)
2. Pure science
3. To draw attention from political failures, for example the Bay of Pigs
4. Economic development of key political states with space jobs (for more on this technocracy argument see Klerkx, 2004; Parker, 2009a).

The first rationale clearly acquiesces with the problematic of America as the transcendental state; it was the explicit emphasis of Kennedy's Moon speech. The second rationale, science, is much less convincing to defend as having any bearing on the lunar landing; indeed even in the early 1960s space scientists were highly sceptical of the scientific justification for Project Apollo; after all human space exploration drains away funding from often more scientifically useful robotic probes and experiments (Burrows, 1998: 323). Secretary of Defence, Robert McNamara, acknowledged this point to Vice-President Johnson, in consultation prior to the speech:

> Major achievements in space contribute to national prestige. This is true even though the scientific, commercial or military value of the undertaking may, by ordinary standards, be marginal or economically unjustified. What the Soviets do and what they are likely to do are therefore matters of great importance from the viewpoint of national prestige (quoted in Beschlosss, 1997: 57).

The problem is, as McNamara and James Webb (NASA Administrator) concluded in a letter to Johnson and Kennedy on May 9th 1961, 'The orbiting of machines is not the same as the orbiting or landing of man. It is man, not merely machines, in space, that captures the imagination of the world' (quoted in Beschloss, 1997: 60).

The third rationale is more convincing (as Beschloss, 1997 suggests); it is surely no coincidence that Kennedy instructed Johnson's consultation on space policy on the 20th April, the day after American warships had evacuated retreating CIA-backed Cuban exiles from a planned Cuban invasion. While it was uncertain how much media or public attention could be drawn away from this and future political failures (by 1961 Kennedy could easily foresee more American military involvement in South East Asia, Africa and Latin America), it was transparently clear how a global audience had reacted to Sputnik. Nevertheless, this rationale cannot fully explain why increased space exploration was the chosen policy solution, let alone the Moon. Why was the Moon the chosen stage for this performance, rather than the production of a fleet of supersonic airliners, even an orbital space station or a fully re-usable spacecraft?

The fourth rationale for the Moon launch seems to be less a reason to go to the Moon than a beneficial after-effect. In that, while NASA jobs and contracts may have been, and indeed still are, of concern to American politicians from both political parties, political support for the Apollo program, as Logsdon (1994) explains, was doomed to be rather ephemeral, not least because, as this chapter has been at pains to make clear, it was staged through a contextually specific political appropriation of the image of America as a transcendental state. Even in 1961, McNamara noted that without Project Apollo there would have been an excess of labour in the aerospace industry, while Kennedy's Secretary of Labour, Arthur Goldberg, seriously doubted its economic value (Beschloss, 1997: 60). Moreover, as Parker (2009a) argues, while Webb may have been seduced more by NASA's technocratic rather than mythological aims, we can only explain the rapid development (and demise) of the former with reference to the latter. Further engagement with this issue is the subject of Chapters 5 and 6.

Kennedy's Moon speech indicates that his vision for outer space gathered its shape and strength not from a series of logical or representable strategic threats presented by the USSR, scientific research questions or even economic or social policies, but rather from the already mentioned set of intangible and un-specific fears, desires and passions that were underpinned by the equally sublime image of America as the transcendental state. It might be tempting simply to read the nature of this sublime threat as simply one of technological prowess (following Nye, 1994); yet there is a significant geographical ingredient to the affective power of Sputnik and Kennedy's Moon speech, drawn from a long-standing reservoir of transcendental motifs incubated across American history, first projected onto outer space by von Braun and Bonestell, and now Webb, Johnson and Kennedy.

After Kennedy's speech an imaginative cosmography became practically, as well as culturally, central to conceptions of American identity. The Moon speech appears to embody *the* moment through which a mythology of writing American into the world as the exceptional universal destiny of humanity, a mythology that had circulated in New World popular geopolitics since the sixteenth century, had come to the fore of American policy decision-making. Borrowing O'Tuathail's (1999: 110) typology, we can evidence a translation from American popular to practical geopolitics. However, the closer this mythology was brought into the policy process the more a tension within it became exposed: how can a bounded spatial entity, a State, with finite institutional capacities, budgetary resources and strategic possibilities, organize and sustain a policy commitment around to what amounts to a messianic claim for transcendental sovereignty, or in Deleuze and Guattari's (1988) parlance, for absolute deterritorializaiton? This tension inflected the development of Project Apollo, and is the focus of the next chapter.

Chapter 3

Placing the Moon

Mythologizing the Moon

President Kennedy's Moon speech to Congress in 1961 unmistakably evoked an image of an American transcendental state; Kennedy's vision for human exploration of the Moon appeared lifted straight from the pages of the *Collier's* series 'Man Will Conquer Space Soon (MCS)!.' Yet to fully understand how this decision was taken in the way it was, and how it developed notions of America as a transcendental state, rather than merely mobilized them, we must go beyond the political arguments framing the decision and explore the significance of the Moon itself. Many space commentators (for example Beschloss, 1997; Burrows, 1998; Cadbury, 2005; Dickson, 2001) offer a persuasive argument that America was best placed, or at least equally placed, in terms of its technology to get to the Moon first. In effect, the race to the Moon offered America a level playing field in space exploration. However, this still does not fully account for why this particular race was so important; why did politicians invest their reputations and budgets in it, how did it capture a global audience, and why did scientists in the US (and the USSR) push themselves often beyond normal practices of rigorous scientific development to win this race,[1] knowingly risking human life in the process?

Many of the reasons given for the political push for lunar landings, over other conceivable space contests, emphasize the role of the race over and above the destination—it was simply a race America *could* win (Burrows, 1998; Cadbury, 2005; Dickson, 2001). Indeed Kennedy himself recognized that this was a race, as revealed in a 1962 conversation with NASA administrator James Webb: 'This is important for political reasons, international political reasons and this is, whether we like it or not, in a sense a race' (Kennedy, 1963). Yet, as explained in the previous

---

1   For example, heated debates occurred within NASA over the use of 'All-Up testing' (AUT), where all Apollo components were to be tested at once together rather than individually. While the latter procedure was the accepted norm by many scientists and engineers to ensure maximize safety and isolate problems, the former was potentially faster. In the event the pressure to use AUT won out as NASA's new head of Manned Space Flight, George Mueller, convinced senior engineers including von Braun to approve AUT in 1964, as Kennedy's 'end of the decade' promise loomed (Cadbury, 2005: 278). The Apollo 1 fire and the tragic deaths of astronauts Ed White, Gus Grissom and Roger Chaffee illustrates some of the pressures for the kind of fast testing and development program that 'All-up testing' exemplified; although after the Apollo 1 fire, von Braun urged more caution in testing, especially for capsule development (Burrows, 1998: 410-1; Cadbury, 2005: 307-10).

chapter, within his Congressional Moon speech of 1961 Kennedy also evoked a powerful image of the value of the Moon to make this race important in the first place: the winner of this race would not just be a leader in space exploration, or science and technology, but something far grander, they could shape the destiny of humankind. In this chapter I ask—how was it that the Moon was able to be constructed as a surrogate place to claim the destiny of humankind?

Kennedy and Johnson were not the first to regard the Moon as an important stepping stone for control of the Earth. In 1958, for example, Air Force Research and Development vice-president Brigadier General Homer Boushey coined the maxim: 'He who controls the Moon, controls the earth' (quoted in McCurdy, 1997: 76). Boushey's remarks, and those of other military leaders at the time (McCurdy, 1997: 64-6), reinforced a tradition of popular stories from scientists (including Wernher von Braun), science fiction writers and military experts that emphasized how control over the Earth could be carried out by controlling outer space. An article published in *Collier's* magazine on 23rd October, 1948, titled *Rocket Blitz*, describes how the Moon, 'could be the world's ideal military base' (Richardson, 1948: 24-5). The article contained images painted by Bonestell of nuclear missiles being fired from the Moon and devastating American cities. But, as McCurdy (1997: 64-7) explains, such stories were based on spurious assumptions that rockets leaving the Moon would require far less energy than those from the Earth. These stories ignored the massive complexities of establishing and maintaining a strike base on the Moon, as well as the argument that weapons technology on Earth could deliver nuclear warheads around the Earth with greater speed and accuracy. It appears that much of the logic framing these stories is tied up with the kind of thinking espoused by nineteenth-century formal geopolitical writing such as that of Halford Mackinder and Alfred Mahan (O'Tuathail, 1996). These imperial scholars described how particular geographical locations are important because of their capacity to control global resources or project military hardware (in actuality, it was eventually a un-fixed location, the transoceanic nuclear submarine, which was to prove most effective strategic space in the age of nuclear weapons—Delezue and Guattari, 1987: 477). Boushey's ideas read like a cosmic appropriation of Mackinder's thesis about the strategic pivot of Central Asia or Mahan's ruminations on the strategic value of different oceanic regions (O'Tuathail, 1996; Sumida, 1999).

While the legacy of formal geopolitics remains evident in contemporary space policy (for example Dolman, 1999, 2002; Oberg, 2007—for a critique Macdonald, 2007), it does not fully account for the kind of space program Kennedy was proposing in 1961. Kennedy was not intending to establish a missile base on the Moon (as suggested in the pages of the *Collier's* MCS series), land armies there, or use it to survey Soviet military sites or act as a military communication relay. Project Apollo was always a thoroughly civilian endeavour. In a speech at Rice University on 12th September, 1962, Kennedy pontificated '[whether space will] become a force for good or ill depends on us, and only if the United States occupies a position of pre-eminence can we decide whether this new ocean will be a sea of peace or a new, terrifying theatre of war' (Kennedy, 1962). This is why it is perhaps more useful to compare the decision to go to the Moon, not to the claiming of American or British imperial possessions through formal geopolitics (contra, Dickson, 2001: 216), but to the peaceful, although militarily organized, Lewis and Clark Expedition that set off to explore, map, and implicitly claim, the American West in the early nineteenth century.

Importantly, the Moon could become the focus of an American project of national aggrandizement because it helped reproduce motifs familiar to accounts of the American West. Crucially in this regard, the Moon offers a landscape. The lunar landscape can be delineated in space and time as an Earthly landscape might; it could be seen to be claimed (especially important given the worldwide proliferation of television in the 1960s and the potential to televise the lunar landings) in exactly the same way as an Earthly territory: physical occupation. By contrast, an orbital plane, even a highly strategically useful one like geo-stationary orbit (Collis, 2009; Dolman, 1999, 2002), exists simply in terms of abstracted spatial co-ordinates and would be far harder to *visibly* claim territorial occupation; it is a Deleuzoguattarian 'smooth space,' more comparable perhaps to the open oceans than land (Burrows, 1998: 332; Steinberg 2001). Perhaps most importantly, the Moon offers a distinctive vantage point from which to observe the Earth. While the Earth can of course be more accurately observed from a spacecraft or satellite orbiting the Earth (on this point see Macdonald, 2007 and also Parks, 2005), the Moon provides a more familiar vantage point. It resembles the magisterial gaze offered by the highest peaks of Rocky Mountains from which, to survey, claim, and command, the American plains, and with it the past, present and future of humanity (Boime, 1991). This lunar vantage point offers a strange, but equally familiar, vantage point from which to claim and control, not just the space, but also the time, of the Earth and its peoples. The Moon has watched over the Earth for billions of years, becoming the center of many spiritual cosmologies. As Cashford (2003: 4) explains, the Moon has long helped us make sense of human consciousness, time and destiny, with its regular cycles of birth, growth, decay and death.

Unsurprisingly, NASA drew upon mythological motifs when naming its spacecraft and missions, many of which have mythological origins that announce an affinity with, usually classically Greek or Roman, deities. The first American human space mission, 'Mercury,' corresponds to the Roman God of travelling and trade: 'the Olympian messenger who was the grandson of Atlas and the son of Zeus' (Burrows, 1998: 287). While 'Apollo' refers to a Greek God, who was as, Karegeannes *et al* (1976) explains, 'god of archery, prophecy, poetry and music, and most significantly he was god of the sun. In his horse-drawn golden chariot, Apollo pulled the sun in its course across the sky each day' (p99). In this sense, Apollo's act of pulling the Sun, though also a reference to the passage of the spacecraft towards the Moon, evokes a sense of Apollo and so America, leading humanity's destiny.[2] Pervading NASA's lunar symbolism is another notion involved in the messianic image of America as the transcendental state, namely the concept of rebirth. Cashmore (2003) explains how the Moon's cyclical transformation has long been associated across numerous spiritual mythologies with rebirth. Similarly, a Judeo-Christian messianic image of rebirth, as salvation, had percolated through American culture since Puritan times (as explored in Chapter 2). Burrows (1998) explains how notions of humanity's rebirth became entangled in the Moon landing:

> For them, as for members of most other religions, purpose transcended both time and the individual and became an immortal crusade. Spaceflight provided the means to undertake the ultimate, endless process of discovery and regeneration for people who took it as an article of faith that the end of exploration would lead to the end of civilization (p332).

The political rhetoric surrounding the inception of Project Apollo directly evokes these messianic mythologies, wherein America will lead humanity's destiny and salvation, recalling Turner's 'Frontier Thesis,' and by extension myths of American Manifest Destiny and Divine Providence.[3] And yet, Kennedy's use of this distinctly American mythology only occurs as a specific political response (as Beschloss, 1997 makes clear), drawn upon, above all, as a geopolitical practice (O'Tuathail, 1996) to essentialize a cosmic order between different peoples and dramas, an order which appeared under immediate threat in the early Cold War era, and especially post-Sputnik. On their own (that is, without the contrast between ideological systems, moral agendas, peoples and dramas) the same said transcendental motifs would surely just as easily dissipate into the sublime ether. Indeed, in the wake of Apollo 11, President Nixon remarked on how the timeless inevitability of space exploration meant it should not be unduly rushed at all costs (Logsdon, 1995: 385-6). President Kennedy himself, as is often noted, was not interested in space travel for its own sake, as is clear from a tape released by Kennedy's presidential library in 2001—Kennedy is heard saying:

> I'm not that interested in space. I think it's good, I think we ought to know about it. But we're talking about fantastic expenditures. We've wrecked our budget, and all these other domestic programs, and the only justification for it, in my

---

2   For more detail about the naming of NASA space missions and space hardware, including mythological connotations see Karegeannes *et al.* (1976).

3   Indeed, Greg Klerkx notes how copies of Turner's 'The Significance of the Frontier in American History' was widely distributed around NASA in the late 1950's (Klerkx, 2004: 152). Meanwhile NASA's Apollo-era administrator, James Webb, similarly drew upon analogies with Turner's 'Frontier Thesis' to emphasize the role America would play in renewing and leading humanity's faith in moral ideals, such as democracy (McDougall, 1986: 387-8).

opinion, is to do it in the time element I'm asking, in other words to win the race to the Moon (Kennedy 1963).

Given such pressures, it became increasingly obvious that any lessening of zeal in the political, religious and cultural exigency for rediscovering an image of America as *the* transcendental state would undermine the rationale for developing such a technically ambitious and costly space program. In other words, NASA's ultimate success in securing America's image of itself as exceptional, as messianic, by conducting lunar landings, would, inevitably, sow the seeds of the demise of the political and public prioritization of space exploration (Burrows, 1998: 383, McCurdy, 1997). Furthermore, NASA's own activities were just as easily capable of inspiring new cosmic mythologies and transcendental thoughts, many of which counteracted the nationalistic imperatives at the heart of its mission. Of particular significance here is the launch on December 21st, 1968, of Apollo 8, the first human journey to orbit the Moon.

Transcendence beyond Borders: Apollo 8

The flight of Apollo 8, the second human mission of the Apollo program, is especially significant in understanding the problematic of America as the transcendental state. The mission was originally intended as an Earth orbit test of the Command-Service Module (CSM) and Lunar Module (LM), but after the US intelligence community reported that the USSR was itself nearing a Moon orbit test, it became a lunar orbit test of the CSM (Burrows, 1998: 418-9). Despite these hurried changes to the mission profile and the astronaut's training program, it was a technical success and paved the way for the Apollo 11 lunar landing. Apollo 8 was also perhaps the most spiritually charged mission of NASA's lunar program: on Christmas Eve 1968 the three astronauts on-board the Apollo 8 Command Module—Bill Anders, Jim Lovell and Frank Borman—recited verses 1-10 of *Genesis* from the King's James version of the Bible, while broadcasting live to an estimated half a billion people worldwide (the world's highest television audience to that date) during their tenth orbit of the Moon (Burrows, 1998: 420). In this act, the passage of their spacecraft around the Moon appeared less as a technical test and more as a step closer the genesis of the cosmos and life on Earth—a step closer to God (on this see Benjamin, 2003: 57; Cosgrove, 2001: 257-62). Apollo 8 was also significant because of the images it relayed back of the Earth rising over the lunar surface, as the capsule passed within seventy-one miles of the lunar surface: the most well-known of which was the image *Earthrise* (Figure 3.1)

This image has regularly been praised for its sublime beauty, its uniqueness, and its capacity to denote the fragility of the Earth and humanity (Burrows, 1998: 420; Benjamin: 2003: 56; Cosgrove, 2001, McDougall, 1985: 412). Viewed from within this analysis of the American transcendental state, this image of *Earthrise* also easily affirms those celebrating America's immanence with God, as it produces

Figure 3.1   *Earthrise*, NASA image: AS8-14-2383
Source: NASA

a Olympian, God-like, gaze of Earth (Cosgrove, 2001). The requirement for such national-divine reconciliation, must have, as McDougall (1985: 411) argues, seemed all the more pressing after the tumultuous events occurring in 1968, from the assassinations of Robert Kennedy and Martin Luther King, to public riots in Chicago, and the TET offensive in Vietnam, which had been an unmitigated and tragic military failure. For Cosgrove (2001: 257), the Olympian gaze of *Earthrise* also evoked a sense of the fragility and wholeness of humanity (and the Earth) adrift in the cosmos; thus perhaps the photograph could also foster hope towards a unified humanity, reborn anew in a world of peace and hope (pre-empting Sagan, 1994). Some commentators sought to reconcile the transcendental and nationalistic qualities of the image and the mission—the *Los Angeles Times* reported, 'The Apollo 8 flight … comes as a welcome talisman of future good fortune—a kind of reassurance that we are still a nation capable of great things' (quoted in McCurdy, 1997: 102). In *Rocket Dream*s, Benjamin (2003) describes the flight of Apollo 8 as a journey to the Divine: 'As purveyors of the new knowledge, the astronauts were aggrandised by the act of travelling upwards, their thoughts at once elevated and rarefied. They'd become privy to the vantage point of God' (p57). For many, the Apollo 8 *Earthrise* photographs reassured an image of the exceptional American affinity with the Divine (Burrows, 1998: 420), and so instilled an American lead in humanity's moral and political destiny.

We can work through some of the sublime effects of Apollo 8 through a brief diversion through Deleuze and Guattari conceptualization of the relationship between transcendence and immanence. Within Deleuze and Guattari's philosophy, the movement and speed of our thoughts across the infinite Cosmos— an absolute deterritorialization—can be elaborated further through their concept of 'pure immanence' (Deleuze and Guattari, 1994: 35-60). This concept of immanence differs greatly from how the sublime has been understood thus far in this study. Crucially, for Deleuze and Guattari (1994), instead of the infinite potential of thought being measured against something else (such as the infinite reaches of Nature/Cosmos or God, as in the Kantian sublime); it is rather thought itself that can contemplate its infinite movement. In Deleuze and Guattari (1994) words: 'the image of thought, that image thought gives itself of what it means to think... Thought demands only movement that can be carried to infinity. What thought claims by right, what it selects, is infinite movement or the movement of the infinite. It is this that constitutes our image of thought' (p37). This image of thought is termed by Deleuze and Guattari (1994) 'pure immanence.' This concept therefore introduces the notion that: 'Whenever immanence is interpreted as immanent *to* Something, we can be sure that this Something reintroduces the transcendent' (Deleuze and Guattari, 1994: 45; original emphasis). So for Nature, God, a State, or the Self, to appear transcendent it must be become possible in some manner to claim that any of these entities is the source and destination of the infinite movement of thought.

Deleuze and Guattari (1994: 35-60) describe how one transcendent being, after another, whether Nature, God or the Self, acts as a substitute, or refuge, for the infinite movement of thought. In the case of the American transcendental state, all manner of other transcendental refuges, including God, Reason/Technology, and Nature are variously harnessed by the State; these processes were described in Chapters 1 and 2. Each of these refuges, or in their terms, 'planes', offers another means for the State to capture the infinite movement of thought, and render itself *seemingly* transcendent and immanent, omnipresent and omnipotent (Deleuze and Guattari, 1994). Deleuze and Guattari (1994) make this point when they describe how 'Whenever there is transcendence... imperial State in the sky or on earth, there is religion' (p43). Yet, as Deleuze and Guattari (1994) are keen to stress, these captures of 'pure immanence', and other transcendental refuges, by the State are only ever partial; they are always subject to leaks, or overflows.

Overflows are evident in many responses to the *Earthrise* photograph. Many commentators see in the *Earthrise* images suggestions of a new mode of transcendence, based upon the immanence of a transcendental humanity, rather than a State. This transcendence appeared as a togetherness or interconnectedness of the fragile and unique human race in the depths of space (Sagan, 1994). Here images of humanity's fetal origins or 'spaceship earth' emerge as popular and potent symbols of humanity's shared destiny (Benjamin, 2003: 59-60; Cosgrove:

2001, 257-62; Ward, 1964; Henry and Taylor, 2009). Astronauts also remarked on this sense of human transcendence, none more so perhaps than the Apollo 14 astronaut Edgar Mitchell, who started the quasi-religious 'knowledge system' called Noetics, after his feelings of cosmic 'interconnectedness' in Space (Benjamin, 2003: 60-1). The mythology scholar and teacher Joseph Campbell describes similarly sentiments when viewing the Apollo 8 image:

> The Earth is a heavenly body, most beautiful of all, and all poetry now is archaic that fails to match the wonder of this view ... Now there is a telling image: this earth, the one oasis in all space, an extraordinary kind of sacred glove, as it were, set apart for the rituals of life; and not simply one part or section of this earth, but the entire globe now a sanctuary, a set-apart Blessed Place (quoted in Dickson, 2001: 221).

The political theorist, William Connolly, similarly explains how a new global political-ecological consciousness could be articulated through the Apollo 8 image: [the distribution of the *Earthrise* image] 'to diverse places in the world... provides a bountiful source of energy for cross-country ecological movements—a new perspective on the world enabled by speed' (Connolly, 2002: 198). Similarly, Henry and Taylor (2009), observe how 'the Apollo images resist being narrated as merely conquered frontier and instead powerfully revealed the Earth's agency as an autonomous, self-regulating biosphere' (p191). Beyond the environmental politics of these observations, what these commentators suggest is that images, like *Earthrise*, as well as the equally famous *Whole Earth* image, taken from Apollo 17, deterritorialize the Earth of territorial differentiation, suggesting our inherent connectedness, our shared humanity (Cosgrove, 2001: 258).

These varied responses all contest the image of the American transcendental state described in earlier chapters. These sublime feelings do so by reconfiguring the relationship between immanence and transcendence. Instead of 'America' being configured as the source and destination of the infinite movement of thought (as in the way God/Reason/Nature are encompassed by America in the Olympian gaze of the Genesis reading Apollo astronauts), it appears that a transcendental, globalized, humanity might become our refuge for the infinite movement of thought. And thus, as space travel expands, and the Earth is reduced to a 'pale blue dot' (Sagan, 1994), America is rendered invisible—deterritorialized—in the vast Cosmos.

America as Humanity: Apollo 11

If Apollo 8 helped to prompt a rethinking of the relationship between space exploration and American nation building, Apollo 11 appears to affirm exactly the opposite: it continues to function as a potent conduit for the apotheosis of America. On launch day, 16th July 1969 at 9:32am EDT, the massive Saturn V rocket carrying the Apollo 11 Command Module (CM), *Columbia*, and the LM,. *Eagle,* departed launch pad 39A in front of a crowd of more than half a million people, packed into the area surrounding the launch complex at Kennedy Space Center, Florida (Nye, 1994: 237-52). Burrows (1998) describes the event as: 'a gaudy celebration of technology and resolve and patriotism that was as close as America has ever come to throwing a spectacular medieval tourney and national circus' (p422). Nye (1994) reads the launch of Apollo 11 as a pilgrimage to worship a sublime shrine of America: 'the pilgrim to Cape Canaveral [the geographic location of Kennedy Space Center] realizes patriotism through the experience of the absolutely great in technology' (p241). Just over four days later, astronauts Edwin 'Buzz' Aldrin and Neil Armstrong landed upon the lunar surface, Armstrong's words captured perfectly the sense that this was a momentous advance for the human race not just spatially but also temporally. As he left the foot of the LM he famously told the world (almost one fifth of the world's population watched on television[4]): 'That's one small step for … man … one giant leap for mankind' (Burrows, 1998: 429). Despite the emphasis upon 'mankind'—in both Armstrong's words, and in a plaque left on the Moon reading 'We came in peace for all mankind'—the Apollo 11 astronauts and the following Apollo missions planted the American flag, not the UN flag, on the Moon to claim this achievement for America (Burrows, 1998: 429). It is difficult to view these conflations between national scientific and engineering (and managerial) success, and the claiming of the destiny of the human race, within Project Apollo, as anything other than the assumed apotheosis of the American state.

On the 24th July at 12:50 EDT the Apollo 11 astronauts were met back on Earth by President Nixon, aboard the recovery ship, *USS Hornet*. Nixon had been inaugurated less than six months previously, and had been highly sceptical of Kennedy's space boosterism during the 1960 Presidential campaign (Hoff, 1997: 96). Notwithstanding Nixon's Eisenhower-like preference for Space pragmatism, he took the opportunity to tell the gathered press and the three astronauts: 'This is the greatest week in the history of the world since Creation' (p430). However, the tide was turning against such cosmic national aggrandizement. By the landing of Apollo 11, the work of NASA was already being scaled back. Even as far back as 1966, Congress had refused to fund a new, evermore ambitious phase of human space exploration, despite the protestations of NASA Administrator, James Webb (Logsdon, 1995: 383). As a result, by August 1968 Webb had started the process of shutting down the production line of Saturn rockets used by Apollo (Logsdon, 1995: 383); yet Johnson had left the most important decisions of NASA's future direction to Nixon and NASA's next Administrators, Thomas Paine and James Fletcher. By 1974 NASA would receive just over a third of the funding that it had in its 1965 peak of $5.25billion (Burrows, 1998: 442).

---

4   The estimated television audience, was 600 million, this was the highest single audience event, since the invention of television, surpassing the Apollo 8 live broadcast on Christmas Eve, 1968 (Burrows, 1968: 427)

The short explanation for this downsizing was the massive amount of funds required to fund Project Apollo. At its 1965-6 highpoint, funding for Apollo was approximately 5.5 percent of a total federal expenditure of $107billion (Burrows, 1998: 425). For comparison, this 5.5 percent equates to approximately $192billion in 2012 US federal funds (11 times greater than NASA's 2012 budget—NASA, 2014). What of course was ultimately really lacking by the late 1960s was public support for ambitious space exploration, and in turn the political currency for a majority of politicians to support such an ambitious program (Burrows, 1998: 433; Hoff, 1997: 92-3; Logsdon, 1997: 210). The downsizing of NASA from the late 1960s has been used to criticize the unsustainable rationale behind the Apollo program (Logsdon, 1969; 1994), not least by Nixon himself (Hoff, 1997). Yet the decision to scale back NASA's funding also clearly demonstrates exactly how Project Apollo was a product of a potent, but precarious, alignment of political ambition, frontier nationalism, quasi-religious transcendentalism, Cold War anxiety and technocratic government and management.

Since the mid-1960s, the White House and Congress had appeared less convinced of the huge cost of continuing the civilian space program on the same ambitious scale, in part because other issues appeared greater threats to national security (than Soviet space dominance), notably the Vietnam War and civil unrest in the US (Burrows, 1998: 441; Logsdon, 1995). If the passion for extending American manifest destiny into Space was in part intended to bolster, or at least exhibit, American security and prosperity, then it now seemed these goals would be better served by a fresh approach to international relations or socio-economic policy. While NASA continued to stress its social and economic value, from its investment in contractor jobs to spinoff inventions, like freeze-dried food (Hoff, 1997; McCurdy, 1993), such arguments appeared to resonate less with an electorate, who by the 1968 Presidential election, appeared more interested in Nixon's promises to check 'runaway government' and curb government expenditure, including the costly military deadlock in Vietnam (Nixon, 1968). Indeed, despite the impending lunar landing, space exploration barely featured in the 1968 Presidential election campaigns (Hoff, 1997).

No doubt many of the reasons for this shift in popularity towards Space correspond to broader changes in public attitudes towards technology and technocratic 'big government' (see Chapter 6). But it is also important to recognize that even for those who believed, post-Sputnik, that America's global leadership was under threat, NASA's very success in Project Apollo dissipated the imperative for such an ambitious space program (Burrows, 1998:432-3): NASA's rather one-sided victory in the race to land a human being on the Moon made it increasingly apparent the USSR could no longer claim global supremacy as it had done in the 1957 (Benjamin, 2003; Burrows, 1998; Dickson, 2001; McCurdy, 1997). By 1968, the Soviet 'menace' simply appeared less immediately threatening, notwithstanding on-going military failures in Vietnam. The peaceful resolution of the Cuban missile crisis, the signing of the limited test-ban treaty in 1962, as well as the UN Outer Space treaty signed in 1967,[5] and the later Strategic Arm Limitation Treaty negotiations, all served to downscale perceptions of the omnipotent, omniscient Soviet threat in outer space that had presaged Project Apollo (Burrows, 1998: 433; McCurdy, 1997).

Epitomizing this shift against seeing Space and the Soviet Union as a threat to American global leadership, in 1974 the final outing for an Apollo spacecraft was as part of project of cooperation between the US and USSR—the 'Apollo-Soyuz Test Project.' In this mission a US Apollo and USSR Soyuz spacecraft docked in Earth orbit, in an act intended to reflect a new spirit of trust and international co-operation between the two nations in outer space and beyond, part of Nixon's wider push for Détente with the USSR. Public and political attitudes had clearly shifted. By the 1970s, as Burrows (1998) concludes, 'civilian space was increasingly seen as being mostly irrelevant to the nation's basic needs' (1998: 441). However, while public, and political, attitudes had changed, political investment in NASA was, as Parker (2009a) suggests, never simply about Space itself, it also rehearsed a specific way of organizing American society on Earth. While some aspects of this attempt at terrestrial organization (or in Deleuzoguattarian parlance 'reterritorialization') were highly commensurate with the myths of cosmic transcendence, many were not. These recursive relationships, between myth and management, transcendence and technocracy, the ideational and technical, constitute the subject of the Chapters 4-6.

---

5 Both the US and the USSR signed this treaty. Within its 27 articles it stated that celestial bodies in outer space could not be claimed as part of the sovereign territory of states party to the treaty and that the same said states were not permitted to place weapons technology of any kind in outer space (UN, 2008).

Chapter 4

Technocracy in the Space Age

Conceptualizing Space Age Technocracy

Numerous commentators (for example Bizony, 2006; Feenberg, 1999; Klerkx, 2004; Launius and McCurdy, 1997; McDougall, 1985; Parker, 2009a) have remarked that the technical success of NASA, exemplified by the organization of Project Apollo, also helped incubate, and validate, a rationalistic, elitist, model of societal organization, better known as 'technocracy'. This model assumes that new technology, scientific and engineering expertise, big business and big government, can gather together around logics of efficiency and cost-effectiveness to solve an array of economic, environmental, social, and even moral problems. Thus, Project Apollo can be read as far more than a mission into Space, it was an experiment in technocracy: 'Whether in decaying cities, outer space or Third World Jungles, American technology would overwhelm the enemies of dignity' (McDougall, 1985: 407). Parker (2009) elaborates this technocratic thesis, quoting James Webb, NASA's Apollo-era Administrator:

> He [Webb] wanted NASA to be an example of large scale interventionism that required 'rapid advances in so many disciplines—engineering, physics, astronomy, mathematics, economics, political science, psychology, public administration—the whole list of the physical behavioural and social science.' He wanted the space program to change the USA (p. 319).

At the heart of technocratic thought is a utopian attitude towards technology. The beginnings of this way of thinking about space technology can be found in Barbara Ward's popular book, *Spaceship Earth*, published in 1966. As an influential international economist on social development, Ward advised Kennedy, McNamara and Johnson. Ward's *Spaceship Earth*, presents a quasi-spiritual celebration of the utopian potential of space technology: ' . . . one of the fundamental moral insights of the Western culture which has now swept over the whole globe is that, against all historical evidence, mankind is not a group of warring tribes, but a single, equal and fraternal community' (Ward, 1966: 148). She elaborates, with reference to new transport technologies including the prospects for space travel: 'now distances are abolished. It is at last possible that our new technological resources, properly deployed, will conquer ancient shortage. Can we not at such a time realize the moral unity of our human experience and make it the basis of a patriotism for the world itself?' (p148).

Ward's (1966) book, along with similar comments by her close friend, US Ambassador to the UN (and two-time Democratic Party Presidential nominee), Adlai Stevenson, pre-empt notions of global togetherness (Stevenson, 1965) that inflected reactions to the Apollo 8 images in December, 1968 (McDougall, 1985:407). The transcendental language of *Spaceship Earth* is wedded to a technocratic vision: technology, and above all American technology, will not only reveal, but form *the* solution to the world's problems. The difficulty is that this model of technocracy threatens democracy because, as Feenberg (1999) puts it, technocracy:

> Represents a generalization to society as a whole of the type of 'neutral' instrumental rationality supposed to characterize the technical sphere. It assumes the existence of technological imperatives that need only be recognized to guide management of society as a system. Whether technocracy is welcomed or abhorred, these deterministic principles leave no room for democracy (p75).

Feenberg (1999) explains how technocracy invariably nurtures an elitist attitude to society; it transfers democratic, political and ethical debate, including debates over technological development itself, to a group of professed 'experts' whose alleged technical expertise enables them to identify and accomplish universal human needs. But technocracy also demands faith by scientists and engineers, and their supporters, that *their* technology (and not that of a rival State—on this point see Nye, 1994: 241) serves, or *determines*, a moral purpose (that is, technocracy breeds 'technological determinism'). Marshall Space Center Director, and early Space booster, Werner von Braun, exemplified this faith.

In 1961 von Braun began to direct the development of the massive Saturn V rocket that would power America to the Moon. Von Braun's own technocratic beliefs were long-established. An ex-Nazi rocket scientist in the Second World War, von Braun is quoted as defending the V-2 rockets that he developed—which would kill thousands of civilians in Allied countries—because he wanted to 'milk the military purse for [his] own ends' (quoted in Burrows, 1998: 78). He later elaborated: 'We were only interested in one thing—the exploration of space ... Our main concern was how to get the most out of the Golden Calf [the Nazi regime]' (quoted in Cadbury, 2005: 10). Von Braun's belief in the moral purpose of rocket technology allowed him to disregard other matters of concern from Allied deaths to working conditions in his factories. After the Second World War, it was revealed, on top of allied deaths, that von Braun had tacitly consented[1] to the use of slave labour in V-2 production, which caused over 20,000 deaths (Cadbury, 2005).

---

1 Primary evidence suggests von Braun fully understood slave labour was being used in V-2 production and moreover that he personally was involved in some of the selection procedures for prisoners (Ward, 2006: 65). Ward (2006) refrains from blaming von Braun directly, as though he was 'aghast at the situation' he 'felt powerless to act' (p64). Yet, it might also be argued that von Braun knowingly consented to develop a highly potent

Figure 4.1   **President Kennedy (first from right), viewing a Saturn V rocket mock-up at Marshall Space Flight Center (Huntsville, Alabama), and meeting the Center Director, Wernher von Braun (second from right), accompanied by vice-President Johnson (third from right) and other guests on September 11th, 1962. NASA Image: MSFC-9801806.**
Source: NASA

In spite of the human cost of his utopian, deterministic attitude towards rocket technology, by the 1960s, after moving to the United States to lead the development of American rocket technology (and thereby avoiding war-crime charges— Cadbury, 2005), von Braun remained convinced that in the long-term his rocket technology was destined to open up Space and fundamentally improve humanity. In his first published writing for the *Collier's* MCS series on the 22nd March 1952,

---

military technology in a political climate from the 1930s onwards which was increasingly militaristic, and where the government had been indiscriminately taking prisoners. Unlike other prominent German rocket scientists, such as Willy Ley, von Braun chose to support this government during the 1930s, even admiring Hitler's 'brilliance' (Ward, 2006: 44), serving in the Luftwaffe, and most importantly choosing to develop a weapon of great destructiveness (p29).

von Braun remarked, 'If we [the United States] do it [explore Space], we can not only preserve the peace but we can take a long step toward uniting mankind' (p179). The challenge, as von Braun, saw it, was urgent: 'Within the next 10 or 15 years, the Earth will have a new companion in the skies, a man-made satellite that could be either the greatest force for peace ever devised, or one of the most terrible weapons of war—depending on who makes and controls it' (p179). Given how integral this technologically deterministic faith appears to the inception of American space exploration, in this chapter I examine how a technocratic mode of social organization became entangled, through Project Apollo, with the image of America as a transcendental state, as described thus far in the study. An especially important element in understanding this recursive relationship is the way in which technocracy promoted efficiency as the only way to organize space (and time) within NASA.

Instrumentalizing Space

The effect of these messianic-apocalyptic, utopian versus dystopia, visions of space exploration, exemplified by von Braun, meant that NASA was essentially put onto a war footing (McDougall, 1985: 381), where, at least in the early 1960s, it could demand almost any support from government to achieve the moral purpose seemingly assumed to be contained within American space technology (Beschloss, 1997). This purposeful action required a great deal of command and control, to enable human individuality and creativity to be aligned in pursuit of a common goal: an American lunar landing before 1st January 1970. But, as Parker (2009a) makes clear these alignments always involved more than just people: 'millions of people and things needed to be brought into alignment and required that managers and engineers developed systems for controlling the times and spaces where these people and things were, and were not' (p326). Thus 'metal, super cold glass, pay cheques and flesh needed to be coordinated, and even that coordination had to made visible, in the shape of tables of numbers, minutes of meetings and flow charts' (Parker, 2009a: 326).

Apollo thus appears orientated towards the production of 'abstract space'—'it is in this space that the world of commodities is deployed, along with all that it entails: accumulation and growth, calculation, planning and programming' (Lefevbre, 1991: 307). This calculable organizational space also corresponds to the 'striated space' of Deleuze and Guattari (1988: 474-5): an instrumental space, where the movements of objects and people can be determined, calculated, mapped and controlled (Dale and Burrell, 2008: 14-16). As Lefevbre (1991) explains, the production of abstract space involves three interwoven elements: Euclidean-Cartesian geometry (where space is conceived as extended, empty and homogeneous, open to logical calculation, as in a map); the dominance of the visual (where objects and subjects are conceived as passive resources); and the phallic formant (where masculine force is perceived and lived as dominating this empty space, as in State bureaucracies such as the police, army etc.). There appears no room in 'abstract space' for ambiguity, for messiness, for wasted time, for unknown space; within NASA all of these complexities were to be translated into calculated risks (Parker, 2009a). Instead space and time within Project Apollo promised to operate like a Fordist clockwork machine, where each moving element contributed to the unity of the whole, all serving a single goal. Central to how this 'conceived' abstract space became 'perceived' and 'lived' (Lefevbre, 1991: 288) in NASA's organizational culture was a particular attitude towards technology (and people): instrumental rationality.

Instrumental rationality, as conceptualized by Max Weber (1978), is a mode of thought and action 'that is determined by expectations as to the behaviour of objects in the environment and of other human beings; these expectations are used as "conditions" or "means" for the attainment of the actor's own rationally pursued and calculated ends' (p24). Thus questions of ethics, values and emotional state are replaced with technical, and economic, calculations of efficiency and cost-effectiveness. Weber (1978) notes that while in the case of technological development instrumental rationality is usually said to serve economic profit-making alone, historically speaking, a part has also played by it in 'the games and cogitations of impractical ideologists, a part by other wordily-interests and all sorts of fantasies, a part by preoccupation with artistic problems, and by various other non-economic motives' (p67). The image of America as a transcendental state is undoubtedly one such non-economic motive. Analyses of NASA culture (as in Parker, 2009a; McDougall, 1985; McCurdy, 1993) describe the prevalence of instrumentalist rationality. This is hardly surprising perhaps given the emphasis in Kennedy's Moon speech on finding the means to a well-defined, if rather quixotic, end: an American lunar landing before 1970.

Interviews with Apollo-era NASA employees confirm how the ends of Apollo were always bracketed from its means; their work appeared a matter of great societal importance as mandated by Kennedy, even if NASA, Congress, or indeed the White House, could not clarify what would happen after Apollo. Apollo-era engineer Thomas Kelly[2] (2000) remembers this sense of destiny in the wake of Kennedy's Moon speech: 'I just accepted the whole thing. I mean, I don't remember questioning any of it. It never entered my mind that we wouldn't get there if we decided to go to the Moon. That was never a concern for me. Indeed, the technology was ready ... ' (p41). Jerry Bostick (2000), an Apollo flight-control officer, describes how, 'You think, hey, the President has established this challenge, the country, by and large, is behind it, it's the most exciting thing that's ever happened, and the money was available. So we just went and did it' (p16). Similarly, Carolyn Hutton, a NASA medical scientist, who worked with astronaut life-support systems during Apollo, and was the first female NASA field-center

---

2 Kelly actually worked for the contractor Grumman not NASA directly. He helped design, test and construct the Lunar Modules for all of the Apollo missions.

director, minimally describes her career achievements in terms of 'accomplishing things and finishing things' (p40).

Kennedy's messianic vision of humanity's future being decided in Space (Kennedy, 1961b, 1962) was reiterated across NASA's organizational culture. Bostick (2000), for instance, describes how he felt after two Gemini spacecraft first docked in 1966:

> ... we didn't really sit around and talk about it. "We've got to beat the Russians" and all that, but we all knew that we were in a race and for once we had beaten them. We kind of waved flags and said "We're proud to be Americans, and, oh, by the way, take that, you Russians" (p20).

These NASA employees work was rendered meaningful in advance by instrumental rationality: the value, or significance, of the Apollo mission was not discussed it was simply assumed (cf. Parker, 2009a: 330). Set against the popular projection of the image of America transcendental state into Space (as described in Chapters 1-3), the ends of American space exploration appeared to be a matter of destiny not debate. As William Rice (2004), an Apollo contract engineer explained:

> The amazing thing about when you would go to a contractor's plant and go down on the floor, go down to the production line, go down and look, you saw people, craftsmen, people who had the skills, the know-how, the expertise. I don't know whether we have that today, but my observation then was that we had the best, and we had a mission, we had an objective and a goal that everybody understood (p5).

At the level of the organization, instrumental rationality produced abstract space: NASA arranged its work into a network of Field Centers, each located to efficiently and cost-effectively contribute to its designated task in the Project Apollo supply chain, just as workers in a Fordist production line. The factory floor for Apollo was America. NASA Headquarters' in Washington DC became a 'center of calculation' (Latour, 1987), where knowledge was received and instructions given. Throughout the 1960s several NASA Field Centers were enrolled in Project Apollo. This contrasts with smaller NASA unmanned missions, such as the Mariner (1962-1973) project of robotic probes to Venus and Mars, in which one Center would dominate the design and operation processes.[3] Some Field Centers had a particularly significant role during the Apollo (and Gemini programs) such as: Johnson Space Center, near Houston (astronaut training and mission control), Marshall Spaceflight Center in Alabama (propulsion systems), the Goddard Spaceflight Center (spacecraft development, scientific payloads and orbital tracking) and the Kennedy Space Center in Florida (launch and assembly).

---

3   The *Mariner* probes were designed by the Jet Propulsion Laboratory, located in Pasadena, California, and managed by the California Institute of Technology. JPL was transferred to NASA by President Eisenhower in 1959.

All of these Centers were established after NASA's inception in 1958 and they constituted the major organizational expansion of NASA during the 1960s (Figures 4.2 and 4.3 below). And moreover, these facilities institutionalized human space flight within the agency.

The creation of the new Field Centers and facilities in the American South (JSC, MSFC, KSC, Stennis and Michoud—all acquired or built after 1960) supported one mission, Apollo, and in particular the construction of the most costly element of Project Apollo, the Saturn V rocket. The decision to appropriate from the Department of Defense the Michoud site (near the New Orleans delta, used by contractors to assemble all Saturn V stages under supervision from MSFC) and construct Stennis Space Centre (on the Pearl River, Mississippi, used to test the first and second Saturn V stages), with access to the Gulf of Mexico, was ostensibly made for logistical efficiency: to allow the fully assembled Saturn V first-stage, which was too large to be easily transported by air, road or rail, to be transported via barges from assembly to testing and onto final assembly at KSC. Similarly, there were technical, and financial, arguments in favor of creating JSC, used for all human space flight training and mission control, at Houston, near to these production facilities. It also made technical and financial sense to appropriate von Braun's army rocket development site in Huntsville, Alabama (to become MSFC) and co-locate the launch facility, KSC, next to a pre-existing Air Force rocket launch based at Cape Canaveral, Florida. Project Apollo thus instilled a spatial division of labour in NASA between the North and West centers (largely those inherited from the NACA, a research led organization) that were primarily dedicated to Space science, Earth science and aeronautic research, and those recently built and NASA appropriated sites in the American South that specialized in the development of human spaceflight, such as astronaut training, large-scale engineering and launch operations. This spatial division of labour in NASA continues to this day.

Many of the prime contractors involved in Apollo maintained offices in close proximity to these centers in the South. For example, Boeing, the prime contractor for the first stage of the Saturn V rocket, established permanent offices near to MSFC, KSC and JSC to liaise quickly with NASA's development teams (Benson and Faherty, 1978). These prime contractors in turn sourced components from hundreds of smaller sub-contractors located across the United States (including Rocketdyne's engines in California); however NASA contracts tended to be tendered to firms located in the states that had NASA field centers. While there were important technical and financial factors influencing the geographical organization of NASA, and in particular Apollo, NASA's procurement practices were hardly beyond political influence. Launius and McCurdy (1997: 241-2) suggest that during the Johnson administration, Apollo engagement and investment in the South, and its blend of Cold War politics, frontier mythology and regional economic growth, was also fully intended helped to soften southern Republicans to future, less-popular big government programs on welfare and security. Thus, despite Kennedy's, Johnson's, and von Braun's, teleological promises of holistic social

Figure 4.2   NASA owned or used field centers and test facilities as of December, 1958 (all locations are approximate)

Figure 4.3   NASA owned or used field centers and test facilities as of December, 1963 (all locations are approximate)

Key to both maps

1. NASA Headquarters (inherited from NACA, 1958)
2. Wallops Flight Facility (shared with Department of Defense, established 1945)
3. Cape Canaveral Launch Facility (shared with DoD, established in 1949) and Kennedy Space Center (KSC) (established in 1963)
4. White Sands Test Facility (shared with DoD, established 1945)
5. Dryden Flight Research Center (inherited from NACA, constructed 1954)
6. Jet Propulsion Laboratory (appropriated from army, 1958, established in 1936)
7. Ames Research Center (inherited from NACA, 1958, constructed 1939)
8. Lewis Research Center (inherited from NACA, 1958, constructed 1940)
9. Langley Research Center (inherited from NACA, 1958, constructed 1917)
10. Johnson Space Center (JSC) (established by NASA, 1961)
11. Michoud Assembly Facility (appropriated from DoD, 1961)
12. Stennis Space Center (SSC) (established by NASA, 1962)
13. Goddard Space Flight Center (GSFC) (established by NASA, 1959)
14. Marshall Space Flight Center (MSFC) (appropriated from army, 1960)

Note. National Advisory Committee on Aeronautics (established 1915)
Source: NASA History Office (2007)

---

betterment directly accrued from the Apollo, as articulated through the rubric of an American transcendental state, the benefits of state investment through NASA were, in practice, highly spatially selective, and profoundly politicized (Bromberg, 1999; Launius and McCurdy 1997; McCurdy, 1993). From its early years the everyday political, organizational and technological effort required to pursue the transcendental ends of space exploration, were already transforming that end. To further unpack how NASA's everyday work, and its technocratic organization, connected to the image of the transcendental state, I now turn towards how work within NASA was managed.

Working in NASA

Project Apollo ushered in a new, more formal, hierarchical, bureaucratic, 'systems' approach to management in NASA (Johnson, 2002; McCurdy, 1993; Parker, 2009a). This 'systems' approach was the result of NASA management's desire to learn from the private sector, especially from its contractors like General Electric and Boeing (Bromberg, 1999: 73-4), but also other corporations like US Steel,

General Motors and Du Pont (McDougall, 1985: 381), as well as the Air Force. Johnson (2002) defines this 'systems' approach as 'a set of organizational structures and processes to rapidly produce a novel but dependable technological artifact within a predictable budget' (p17). Under this approach, corporate contractors were charged with completing projects, and delivering them on time and in budget, in the best way they could. NASA then project managed these contracts through a system of standardized and high-speed bureaucratic flows of knowledge, from Field Center Directors to specialist Program Offices and ultimately the Office of the Administrator, all supporting the direction and approval of designs and monitoring of quality (Johnson, 2002; McCurdy, 1993). Thus, as Bromberg (1999: 61) clarifies, NASA was required to not only match technical expertise in the contractor but recruit people whose quest for detailed quality control exceeded those of the contractor. This style of management contrasted with early engineering management strategies prevalent in the US and beyond, where government had tended to react either retrospectively to encourage sporadic technical development (such as in the railroad, or airplane industries), or to provide only vague objectives for innovation with little direct monitoring and management, such as in NASA's predecessor, the NACA (Johnson, 2002: 2). NASA's approach in Project Apollo by contrast owed more to the bureaucratic scientific management approach, popularized across the world in the early twentieth century by the American mechanical engineer, Frederick Winslow Taylor (Taylor, 1911). Taylorism, as it was later coined, was previously predominantly applied to relatively low-tech, stable production contexts, such as automotive manufacturing; its extension into dynamic research and development only began in 1950s within organizations like the US Air Force and Boeing (Johnson, 2002: 47-80).

Within the production facilities of NASA and its contractors, production lines were organized to efficiently and cost-effectively (and for the contractors, profitably) process Apollo parts. Here space was conceived, perceived and lived as instrumental—new production and launch facilities, as well as spacecraft parts, were designed within strict geometric calculations (Lefevbre, 1991: 289), to maximize a trio of objectives: time efficiency, cost-effectiveness and quality (the so-called 'iron triangle' of project management). Indeed a great deal of this production work involved rigorous quality testing (Bromberg, 1999: 61); this is hardly surprising given the mission, and safety, critical nature of many of the parts (Figure 4.4).

To ensure quality control, NASA gradually developed a standardized system of component tracking or surveillance, mediated by an elaborate chain of paper forms (later computer networks), which tracked the spatio-temporal movements of individual components from their production by sub-contractors right up to launch (McCurdy: 1993: 59-60). This system was termed 'configuration control.' This bureaucratic system ensured that design, cost and progress could be monitored and tracked at every step of the technology's development and production (during Apollo these two stages frequently overlapped—Bromberg, 1999: 72);

Figure 4.4   Workers on a production line inspecting the second stage of engines that would be later used in the Saturn V rocket at Rocketdyne's (later Rockwell's) Canoga Park, California, production facility, 1st January 1960, NASA Image: MSFC-9801772

Source: NASA

thus allowing NASA project managers and senior management to be constantly informed of any necessary design changes and problems (Johnson, 2002: 61-4).

One of the key proponents of such practices within NASA was Col. Samuel C. Phillips who was hired by NASA in 1963 from the Air Force's Minuteman missile program. In May 1964, Phillips issued the *Apollo Configuration Management Manual* to all NASA field centers, which required the establishment of configuration control board meetings to standardize the monitoring of project design, costs and time criteria. Various control processes were implemented by these boards, including a change control process requiring that groups could only make design changes without higher authorization if those changes only affected components that group held responsibility over; this system thus demanded significant technical knowledge of the interfaces between components (Johnson, 2002: 138-41). Parallel systems were introduced to ensure that managers could track production, including a requirement by all parties to use PERT (Program Evaluation and Review) charts and CPM (Critical Path Method), to sequence and monitor the co-ordination of inter-related work tasks. PERT and CME allowed engineering teams in contractors to ensure that delays were avoided by concentrating their resources on those areas of production which were essential for other teams to proceed. When problems were encountered inevitably the proposed solution was to demand increased monitoring. For example, when in spring 1966, the lunar module manufacturer Grumman experienced schedule delays and growing costs, NASA was quick to sack the project manager, create a program control office and even move Grumman's vice president to the factory floor (Johnson, 2002: 145-6).

Charles Bingham, recounts his experiences of working for NASA's management analysis team at JSC during Apollo:

> I was much more interested in getting my fingers into everything, and so the management analysis shop looked like that kind of an opportunity. I ended up doing a whole lot of organization work, not only in the administrative organization, but throughout the whole center. I would work with the heads of the engineering, technical, program management staffs when they had organization problems. Also I would try to develop methods, procedures, business practices, especially those that cut across the operations of the whole Center, all the way from a mail system and design of forms, up to major management systems which I would help design and develop the written procedures to govern them (Bingham, 2000: 6).

Amongst the procedures Bingham and his colleagues helped formulate and implement was a system of 'cost-plus' incentive contracts for production. These contracts, as Bingham (2000) recalls, revolved around an 'award fee' premised on the notion that:

> We will pay all of your costs, but your fee will essentially be awarded based on the excellence of your management performance. That could be stated in a series of criteria—cost, schedule adherence, technical excellence, things of that kind. The intent there, the contractor could earn a larger fee, but only by demonstrating superior managerial effort within the company on the NASA contract (p12).

Such incentives (and disincentives) were used as a tool to further circumscribe thinking around issues of cost, reliability, technical efficiency and timing.

Techniques such as PERT, CPM, configuration control and 'cost plus', increased the visibility, geometric calculability (and the force of the American State) in NASA's workspaces, and beyond; thus systems management produced abstract space in exactly the manner outlined by Lefevbre (1991: 285-88). NASA's workgroups were now not *simply* able to informally feel their way through the design of a component or specification for a project, as had occurred in programs as recent as Mercury (Johnson, 2002: 118). Yet while it might be suspected that such spaces were destructive to individual well-being (following Lefevbre, 1991:

289), many employees' express great passion, and excitement, towards their work on Apollo, alongside moments of informality. Thomas Kelly recollects his time managing the production of the Apollo lunar module:

> ... it was just a lot of fun. We had sessions where we'd get up on the blackboard and sketch different ways of doing things. I remember one session that went on pretty much all day where we looked at different ways of positioning the attitude-control thrusters and finally worked out the arrangement that we used. One of the main things that made the design change from the proposal was the need to get reliability, and we did that by adding redundancy where we could, and where we couldn't, by making the system just as simple as we possibly could. (Kelly, 2000: 8).

A senior Apollo power-generation engineer, William Rice, recalls his reasons for joining NASA in 1962: 'Well because John Kennedy had announced that we were going to put a man on the Moon by the end of the decade and return him safely to Earth, and that just sounded like an exciting thing to be part of' (Rice 2004: 3).

Notions of 'flexibility', 'creativity', even 'ethics' and 'democracy', were not entirely absent within Apollo systems management, but were reformatted within a discourse of instrumental rationality, of instrumental space. Bingham (2000), for example, describes the flexibility of the procurement system NASA developed:

> ... a lot of the problems initially occurred in this arena [procurement]: how do you make the Manned Spacecraft Center function at a very high level of competence and quite rapidly and avoid all the bureaucracy that had overburdened the Department of Defense? There was tons of work to do in this area. So the initial effort was to design a procurement system which was as flexible and as open as possible and could be operated with some rapidity. I got involved in that simply because it was a systems design, and the procurement people were having a lot of discomfort trying that. (Bingham, 2000: 7)

By the same token, Bingham (2000) instrumentalizes the ethics of NASA engineering:

> If you ask an engineer, "How do you decide what a good job is?" or, "How motivated are you to do a good job?" almost every engineer will talk to you about the ethics of the profession. Starting from college on up, when you learn engineering, any kind of engineering, what is built into that is the clear ethic that you will do the best job you can, that you need to pay extraordinary attention to the work to avoid failure. You need to *do things right*, as opposed to cheap or shoddy quality, as a characteristic of engineering work. Part of the ethics of engineering, don't be a cheap, shoddy engineer. (Bingham, 2000: 34; emphasis added).

Even 'democracy' became just another part in the NASA machine:

> There was a great deal of democracy in the management … Everybody was free
> to state their feeling. No one was treated any differently if he objected to what
> management would think than if he praised what management would think.
> Management didn't look for praise. They looked for anybody with good advice
> (anonymous Apollo engineer quoted by McCurdy (1993: 65).

But crucially, as Parker (2009) reports in his analysis of Webb's 1969 book, *Space Age Management* (Webb, 1969b), this instrumental rationality appeared far less appropriate within the upper echelons of NASA, where individuals like Webb (and even von Braun), possessed considerably more freedom to comment on, and influence, the substantive direction of the organization, and nation. Parker (2009) paraphrases Webb's view of NASA's executives: '[they] are not the sort of vulnerable children who need to be looked after by personnel managers, or given instructions by their superiors' (p321). As Parker (2009) explains, this elitism was legitimized through a technocratic discourse of wider public responsibility. Indeed for Webb, the audience for this experiment in technocracy always extended beyond NASA:

> The nations of the world, seeking a basis for their own futures, continually pass
> judgement on our ability as a nation to make decisions, to concentrate effort, to
> manage vast and complex technological programs in our own interests. It is not
> too much to say that in many ways the viability of representative government
> and of the free enterprise system in a period of revolutionary changes based on
> science and technology is being tested in space (Webb, 1963).[4]

Beyond Space Age Technocracy

After the initial impetus of Webb, Phillips, and others, technocratic systems management continued to advance through NASA. Even as budgets were

---

4   What of course was absent from Webb's speech in 1963, was, as Stephen Johnson (2002: 27-9) reminds us, that many of the prescriptive techniques and ideas of this pro-technocratic, centralized and de-politicised approach to technological development and the messianic mode of social organization it nurtured, were in fact developed partly by the Pentagon from von Braun's V-2 production team from their work at Peenemunde under the Nazi regime. McCurdy (1993) similarly explains of NASA's centralized organization: 'The project management techniques were invented before NASA began, to coordinate high technology activities such as the development of military ballistic missile systems. The approach operated on the principle that responsibility for the timely completion of a project could be concentrated in a single management team that would coordinate all parties making contributions to the undertaking' (p47).

tightened after Apollo (Hoff, 1997): the space agency increased its administrative staff from 5 percent in 1961 to 18 percent by 1991 (McCurdy, 1993: 116). This bureaucratic shift was largely because NASA had contracted out more and more work to private companies, including aspects of its administration, in an effort to decrease production costs. In turn, NASA lost some of its in-house technical expertise and skill (McCurdy, 1993: 116-6). Some view this shift away from a perceived balance of administration and technical expertise, within the original version of systems management, as responsible for the downturn in NASA's achievements and ambition after Apollo. McCurdy (1993) quotes one NASA Field Center director who laments 'the politicization of the organization' which led to 'bureaucratic decay' (p109). Bingham (2000) similarly laments NASA's post-Apollo bureaucratic malaise: 'There's another element that's not as easily recognized, but during the time I was in NASA, let's say in the sixties, up to the late sixties, NASA had this reputation for being a quick, hard-hitting, simple, straightforward organization, [but] bureaucratic red tape again built up' (p29).

Such comments suggest that America has not returned to the Moon since Apollo, or achieved the kind of expansive vision of space travel set out in the pages of *Collier's* MCS series, because bureaucracy, and by extension technocracy, has constrained innovation; but the American public long ago grew tired, even bored, of Space, while NASA's bureaucracy had successfully challenged the principal driver for the emergence of that bureaucracy: the Soviet (technocratic) threat to the image of America as a transcendental state. The solution for some to this predicament remains not less bureaucracy, but better bureaucracy, more Apollo-era systems management: if only NASA can demonstrate to Congress it can be more efficient and cost-effective, and re-discover managerial excellence, then it can achieve any end (Johnson, 2002: 228). The influence of NASA's Apollo-era system management on contemporary project management (for example Johnson, 2013), stands as testament to how, what amounts to, technocracy continues to appeal in a liberal democracy as an explicit guide for how people, technologies, and capital, might be aligned, and transformed, to serve spectacular (if elitist) ends, and an implicit guide for how 'ethics' and 'democracy' can be read as synonyms for commitment to, and open-ended debate on, efficiency and cost-effectiveness, and 'politics' as merely a pernicious distraction (on the technocratic legacy of de-politicization in contemporary project management see Hodgson and Cicmil, 2006).

Unsurprisingly, the success of Apollo prompted Congress to search for wider lessons for other government projects (Beniger, 1986). As a result, many of the systems management practices implemented by NASA, soon spread into other areas of US government. For example, NASA developed lucrative support networks within the American university system, especially at leading academic institutions such as MIT and Caltech to cultivate local economic growth (McDougall, 1985: 384-5; Bromberg, 1999: 62). These multi-million-dollar programs led to the construction of space-science and engineering laboratories in many universities, as well as first-hand contact with aspects of system management (McDougall, 1985: 385-6). The aim of these programs was to expand a technocratic 'university-industrial-government' complex (McDougall, 1985: 381-7). Webb also hired executives, including two subsequent NASA administrators, Thomas Paine (1968-70) and James Beggs (1981-85), from large, American, corporations that had undertaken NASA contracts. Webb was convinced that NASA's blend of democratic accountability and legitimacy, industrial attention to efficiency and cost-effectiveness, high-technology and academic creativity, would operate as a technocratic template to address an array of social, economic, technical and environmental problems, and evidence America's global leadership (Bizony, 2006; Parker, 2009a). McDougall (1986) concurs: 'Few challenged any more the notion of state responsibility for directing progress in science, technology and education, for setting social priorities, and forging technical tools to achieve them. Left and Right, hawk and dove, by 1964, most Americans had opted for technocracy ... even as the space program was called into question, the new mode of governance for which it served as symbol was not' (p389). Similarly, Johnson (2002) presents technocratic systems management as part of a messianic narrative of American global leadership:

> The offices, managers, engineers and scientists who created systems management in the first two decades of the Cold War did so because they believed in the efficacy of technology to protect and promote the values of the United States. After this time, the apparent effectiveness of these methods in creating missile, space and computing technologies led technologists and managers in other nations to mimic Americans. Through the combined efforts of these groups of people, technological innovation has become a standardized commodity through the Western world (p230).

Johnson's (2002: 230) insertion of systems management within the story of the American technological sublime, described by Nye (1994), where American space age management instils faith in American global leadership and exceptionalism, is telling. Project Apollo, NASA's technocratic triumph, was always bound up with something beyond itself: transcendental spaces and times which are far from calculable in technical abstractions—sublime mythologies of American exceptionalism (recall in Deleuze and Guattari's parlance 'reterritorialization' always involves 'deterritorialization'). Some philosophers of science even suggest a quasi-religious aspect to our relationship with technology: 'Our god is the machine, the technical object, which stresses our mastery of our surroundings' (Serres and Latour, 1995: 141). This point has important policy implications too: the success of Project Apollo was far from inevitable; it was not fully calculable or predictable and thus easily repeatable elsewhere (in NASA or beyond).

The swirl of transcendental myths about Space in American culture, especially in the late 1950s and early 1960s, while seldom acknowledged by Apollo employees in their experiences of work, are unavoidable in explaining not just individual employee's passions for Apollo, and the commitment evident to its undeniably technocratic forms of organization, but indeed the existence of Apollo. However, equally, the transcendental image of America as a homogeneous, spaceless and timeless, State, was challenged by NASA's mundane organization as it channeled billions of dollars of taxes into the hands of a geographically and socially exclusive collection of corporations and universities. After all, as Lefevbre (1991: 287-9) reminds us, abstract space only *aims* for homogeneity in order to mask the perceived and lived spatial inclusions and exclusions of capitalist prosperity. As Apollo progressed, this assemblage of mythology and management, transcendence and technocracy, the ideational and the technical, was thrown into sharp relief, along with its power laden consequences. One of the more vivid examples of its power effects is the gendering of American space exploration. In the next chapter I will address this subject to illustrate some of the more pernicious consequences of this association between nationalist myth making and space age management. And in so doing examine in more detail the third aspect of Lefebvre's (1991: 287) 'geometric-visual-phallic' triad for the production of abstract space: the 'phallic'—masculine violence.

Chapter 5

Whose Body for Whose Future?[1]

Gender, Technocracy and Technology

As explained in the previous chapter, the technocratic orientation of NASA, instilled by Project Apollo, was reproduced as much through the disciplining of employees bodies and components in production facilities, as in grandiose speeches about the university-government-industrial complex. This is hardly surprising: all Taylorist forms of management, systems management included, seek to train the bodies of workers to move efficiently in space and time, alongside other bodies and material artefacts, in the interests of the organization (Clegg et al., 2006: 53). The form of technocracy ushered in by Project Apollo moves this Taylorist vision further by constructing, and linking together, all manner of novel abstract, instrumental spaces (the contractor production facility, the university laboratory, the testing facility, the launchpad, the spacecraft, Earth's orbit, the lunar landscape), which not only enact this discipline (as in the Taylorist factory), but actively seek to develop new means to optimize the interface between humans and machines, ostensibly in order to improve humanity. When coupled to the image of the transcendental state, this Taylorist agenda not only reifies political assumptions about how bodies should behave but what kind of bodies, and bodily competencies, are useful and desirable in an idealized future.

Within this chapter I explore some of the consequences of NASA's bodily prescriptions with reference to the gendering of American space exploration, starting with the Apollo-era. However, from the outset I must be clear that I do not wish to overstate the capacity of bodies to be disciplined in NASA or elsewhere. As Latour (2005) proposes, individuals, and indeed technologies, seldom function as mere intermediaries of grand social projects, but more frequently mediators capable of modifying, transforming and re-negotiating those projects. At the level of an organization, NASA's own socio-spatial division of labour, as discussed in the previous chapter, has indicated some possible transformations of the image of America as a transcendental state. In continuing to flesh out these analytical trajectories, in this chapter I propose that attention to how bodily discipline is performed through gender in NASA offers more insights into how individuals negotiate technocratic discipline, along with the image of an American transcendental state.

---

1 Material in this chapter was first published in 'Giant Leaps and Forgotten Steps: NASA and the Performance of Gender,' Sage, D, (2009), *Sociological Review* (Wiley), reprinted by permission of the publisher (John Wiley & Sons, http://www.wiley.co.uk).

Contemporary studies of gender and technology (Hird, 2004; Lerman *et al* 2004; McGaw, 1987) also offer a rich starting point to the analysis I pursue in this chapter. This body of work draws attention to the 'mutual shaping' (Lerman *et al*, 2004: 430) of gender and technology.[2] Of particular relevance to the question of how bodies are gendered within NASA is the way in which, as Lerman *et al* (2004) explain, there is a naturalized association between the male body and technological prowess in modern industrial society:

> … … a society that values technological change camouflages the privileges accorded men; they are labelled privileges of technological knowledge rather than of masculinity. The gendered production/consumption dichotomy, so common in industrial society, labels male-coded activities "production" and camouflages work and the technological content of activities labelled "consumption" or "reproduction". Consumption and reproduction activities are generally unpaid or under-paid, invisible knowledge and invisible contributions in a capitalist economy (p7).

Thus, as Lerman *et al* (2004) surmise, gender and technology operate to essentialize each other, through connections so naturalized they are 'hidden from analysis' (p6).

Exceptional People

Connections between space technology and gender are found across the pages of McCurdy's (1993) study of NASA's organizational culture: *Inside NASA*. While McCurdy (1993) omits any explicit discussion of gender from his analysis, his study focused on how the 'norms, customs, values and language' (p4) of NASA's organizational culture reinforce judgments about what kinds of practices, knowledge and technologies could contribute successfully to the organization. McCurdy (1993) describes how the perception that NASA *had* to recruit 'exceptional people' (p50) precipitated a belief that 'Within the NASA culture, exceptional performance and the maintenance of technical capability depended for the first thirty years upon the agency's ability to recruit outstanding people' (p50).

2　Such work is partly a response to oversights by historians of technology, whom as Lerman *et al* (2004) suggest, have 'been sluggish in their attention to masculinity as a crucial critical dimension of the material they have traditionally studied and reluctant to challenge familiar taxonomies' (p426). Similarly, contemporary gender historians are lamented because they 'tend to leave technological issues unexamined, neatly relegated to the "black box" that scholars of technology have been at pains to open' (p426). That said, Lerman et al (2004) recognize the insights of feminist sociologists of science, although they decry their focus upon periods of rapid technical change rather than 'slower paced transformations in gender ideologies over time' (p425-6)

The gendering dynamics of this policy is hinted at by McCurdy (1993) when he writes:

> NASA officials assumed that the success of the civilian space program could be traced to their ability to recruit exceptional people ... Accordingly, they placed demands on their work force that they could not have made if the work force was merely adequate. From outstanding people they demanded outstanding performance. In particular, NASA management demanded that their employees work harder than average civil servants and pay more attention to detail than was expected in a typical government organization. NASA employees, believing themselves to be part of an exceptional group, responded positively. In this way, management's faith in the quality of its people was affirmed. Higher expectations, as a consequence, helped NASA officials overcome many of the obstacles that deter high performance in government agencies (p50-1).

There are clear echoes of these sentiments in Jerry Bostick's (2000) account of NASA culture at Johnson Space Center, during Apollo:

> I learned at both headquarters and at Grumman, what I hadn't realized until I got in those environments was that at JSC in the sixties I was surrounded by competent people. It wasn't just a few like Gilruth and Kraft and Dunseith and Lunney and Charlesworth and Faget. I mean, yes, they're the standouts. Everybody was competent. The teamwork that we had was just incredible (p77).

NASA's senior management reinforced this spirit of exceptionalism within institutional procedures designed to maximize efficiency. James Webb recalls his views on the subject: 'My attitude was if you have 200 project managers who were adequate, you should still remove 10 percent of them every year to keep the pressure on them to do better' (Webb, 1969a). Predictably, very long working hours were the norm despite a salary well below that of industry, especially during the 1960s (McCurdy, 1993). McCurdy (1993) quotes one unnamed Apollo-era NASA employee: 'We just worked sixty, seventy hours a week for most of those years' (p56). The same official then recalled the words of Keith Glennan, NASA's first administrator: 'He said, you have got to go at this program as if you were fighting a war. And he said, this is no place for tired men' (McCurdy, 1993: 56). Another NASA executive elaborates further: 'The space program is a young man's game'; he recalls asking his successor in the recruitment interview: 'This is a rough job. Do you think you've got the stamina to handle it?' (McCurdy, 1993: 57). NASA engineer William Rice recalls similarly experiences of Project Apollo:

> Those were the glorious days of NASA. I look back on that time in NASA and the competency of the people, the drive that they had, the sacrifices that they made of personal time. I mean, it was hard. We were working twelve, fourteen hours a day, and sometimes seven days a week to meet schedules, to stay on

time. But I wouldn't take for the experience. It was great. It was great being part
of the team, it was great working with guys (Rice, 2004: 36).

Many NASA employees, during Apollo, were as McCurdy (1993: 56-60) reports,
young people, overwhelmingly men, living in hotels and motels and working long
hours well into the evening and seeing their friends and family perhaps only one
day a week. Bingley (2000) describes how Johnson Space Center:

> had the opportunity to select very high-quality people to come work on the staff,
> which means that most of them were very highly self-motivated. It was not a
> question of people in dull, sluggish jobs, and you have to try to pump them up
> in order to get them to get away from the water cooler. It was not like that at all.
> Almost without exception, the people who came to work there were just charged
> with this emotion (p23).

This intensive work ethic was supported by the claim of supposed technical
necessity to achieve success not just in Space but ultimately on what McNamara
and Webb, famously referred to as the 'fluid front of the Cold War' (McNamara
and Webb, 1961); this work ethic endured long after Project Apollo. And indeed
aspects of it, such as an 'up or cut' work culture have since become synonymous
with the excesses of contemporary hyper-masculine, intensive work environments,
such as financial services (Grey, 2003). As a gendered performance, this work
ethic presupposed that NASA required people who would dedicate themselves
solely to the space program, exceptional people whose lives exemplified the
disciplinary regime required to realize America's moral destiny, in other words
the personification of Kennedy's famous mantra of '... ask not what your country
can do for you, but what you can do for your country' (Kennedy, 1961c: web
source). Indeed, while concluding his 1961 'Moon' speech to Congress, Kennedy
emphasized the character of NASA's work ethic in exactly these terms:

> It [Project Apollo] means a degree of dedication, organization and *discipline*
> which have not always characterized our research and development efforts. It
> means we cannot afford undue work stoppages ... [he then asks that] every
> scientist, every engineer, every serviceman, every technician, contractor, and
> civil servant gives his personal pledge that this nation will move forward, with
> the full speed of freedom, in the exciting adventure of space (Kennedy, 1961b;
> emphasis added).

These sentiments were frequently reiterated by NASA's senior management.
For example, Deputy Administrator Hugh Dryden is quoted in NASA's *Future
Program Task Force Document* (NASA, 1965) stating: 'We must not delude
ourselves or the nation with any thought that leadership in this fast-moving age
can be maintained with anything less than determined, whole-hearted sustained
effort' (NASA 1965: 19). By conflating American universal leadership with highly

specific work practices, technical strategies, knowledge and techniques, both Kennedy and Dryden presented the values which his image of a 'better' humanity would not just desire but inevitably require.

In this manner, NASA's demarcation of the 'expert body' hidden within the recruitment policy of 'exceptional people' tacitly constructed different bodily practices, knowledges, spaces and technologies, including those in the home or even mall, as somehow perhaps more 'passive', 'excessive', 'banal' or 'marginal' to this exciting epicenter of American modernity. Feminist scholars of technology and science provide some important insights into the essentializing alignment of gender and technology across different spaces and times (compare with, Hird, 2004; Lerman, *et al* 2004). McGraw's (2004), for example, shows how, after the Second World War, American domestic technologies, serving 'mere' functions of food, clothing or shelter were feminized, and thus rendered invisible within more epic, and masculine, American stories of 'human creativity' or 'progress' (p32). In an influential article, the social historian, Ruth Schwartz Cowan, considers how domestic technologies have been systematically excluded from the history of technology, by evoking naturalized gendered distinctions between productive technologies as 'creative' and 'productive' and consumptive/ reproductive technologies as 'banal' or 'routine'. This in turn feeds into the gendered social construction of public and private spaces as respectively 'active' and 'passive' (Cowan, 1976 and also Pacey, 1999: 147-69). More recently, Kevles (2003) characterizes this division as an extension of the frontline of Cold War geopolitics into the American home, wherein 'The soviet's promised women equal opportunities in careers like medicine and engineering; capitalists offered women consumer goods and the luxury of remaining at home to use them' (p3). Presumably for these 'capitalist women' freedom was to be equated not with equal opportunities to work but with freedom from the exhaustion of domestic work, a freedom nevertheless bound up with domestic space. And yet, as Kevles (2003) acknowledges, labour shortages after the Second World War and less time-consuming domestic responsibilities, as a result of household technologies, increasingly meant that women desired and were sought after in a variety of conventionally 'masculine' careers, including NASA.

Undeniably, NASA was never solely a male employer. By the mid-1960s it employed thousands of women, as had its predecessor organization, the NACA. This work included seamstresses working on space suits to aerodynamics engineers and in—flight physiologists (for more information see Moule and Shayler, 2003: 92-107). Moreover, women were often placed in key positions of responsibility such as, for example, computing the flight trajectory for the first manned Apollo (Apollo 8) mission to orbit the Moon and return to Earth (Moule and Shayler, 2003: 97). Indeed, in a December 1961 advertisement in *American Girl* magazine, President Kenney sought to promote women working at NASA, stating:

> In our many endeavours for a lasting peace, America's space program has a new
> and critical importance. The skills and imagination of our young men and women

are not only welcome but urgently sought in this vital area. I know they will meet this challenge to them and to the nation with vigour and resourcefulness. (quoted in Moule and Shayler, 2003: 92)

However, to argue that because NASA recruited women negates its complicity in the masculine the gendering of American modernity, is far too simplistic. Indeed, women had by the 1960s long contributed to American aeronautical science and engineering, not least during World War Two, while essentialist assumptions about masculine and feminine gender roles continued to feed into the importance attached to particular embodied practices, knowledge and technologies across America, not least in NASA. Undoubtedly the presence of women in such stereotypically 'masculine' roles did (eventually) offer a repository of transgressive case studies through which to challenge the basis of essentialist gender distinctions (Sage, 2009). And yet, far more frequently, NASA's portrayal as an exceptional organization composed of exceptional people—the leading vanguard of an American transcendental state—seamlessly corroborated a patriarchal version of social relations. This is because NASA's employees self-disciplined their identities, and bodily actions, to correspond to a kaleidoscope of performed traits readily associated with 'hegemonic masculinity'[3] (compare with, Beynon, 2002: 16-17; Gutterman, 2001: 56-71). In this analytical vein, McCurdy's (1993) evaluation of NASA's organizational culture, and Johnson's (2002) analysis of NASA's systems management, can also both be read as stories of hegemonic masculinity (Connell, 1995): traits such as *risk taking, frontier exploration, control, technical decision making, competition* and *attention to detail* (all associated with hegemonic masculinity—see Connell, 1995) appear in these studies as moral virtues, part of a golden age of American space exploration in which American global leadership was assured.

Plenty of evidence suggests that neither McCurdy (1993) nor Johnson (2002) are incorrect in their assessment of NASA's organizational culture, especially during Apollo. Webb describes how NASA's senior management '… were risk takers and had to be to get the job done' (Webb, 1969a). Eugene Kratz, director of Mission Control at JSC (1964-74), proposed that the core values of NASA were 'commitment, teamwork, discipline, morale, tough[ness], competent[ce],

---

3   The phrase 'hegemonic masculinity' was coined by R.W.Connell in *Masculinities* (Connell, 1995). Connell's thesis, following Butler (1990), is that masculinities are always performed, never given, and are therefore plural and emergent rather than singular and natural. Yet, despite this more constructivist approach, Connell identifies a series of normative traits commonly associated with men's bodies, such as risk-taking, physical prowess, competitiveness, domination and control; these varied (sometimes antagonistic) traits are for Connell regularly associated with 'manliness' and thus compromise multiple, though hegemonic, masculinities. As Connell explains hegemonic masculinities are united being orientated to exclude novel masculinities, and also render other subjects passive and neutral; including, as this chapter demonstrates nonhumans.

risk, sacrifice' (Kratz, 1999). This prevalence of hegemonic masculinity with technocratic organizations can be explained with reference to Arnold Pacey's (1999) book, *Meaning in Technology*. Pacey (1999) explains how expectations that men will become the highest earners are readily coupled to desires to find 'jobs that are technologically interesting or involve powerful machines or some risk tasking' (p155), especially if these jobs pay more, as indeed they often do.

Given the ubiquity of this hegemonic masculine culture within NASA during the 1960s it is perhaps not surprising to find how Charles Bingham, who once worked for human resource development at JSC, expressed the pessimism felt by women (and ethnic minorities) towards employment in NASA:

> If you know NASA at all, you know this is not where women and minorities would normally turn as a first opportunity for a job. At that time [during Apollo] particularly even with the best women in the world, there were not that many women taking advanced engineering programs. That's not to say that they were not out there, but it is to say that you had to work harder to go find them or to make the fact known that Houston was a good place for women and minorities to work. A lot of them didn't believe it. A lot of them didn't believe that you could go into an old-fashioned engineering shop and ever be given any responsibility or become a real partner in the organization. (Bingham, 2000: 14).

Heroism and its Other

One of the most elucidating performances of this rehearsal of masculinity in NASA occurred across the astronaut program. A key conduit for this performance was risk. Beynon (2002) suggests that the performance of acceptable levels of risk often functions as a cipher for the performance of masculine identities. McCurdy (1993), quotes one Apollo astronaut: 'Recognition of risk is what made us as good as we were' (p62), while another states 'But if it [risk of death] was like, one in one hundred, you would do it, you take it ... There were so many ways it could happen' (p63). Across such statements astronauts fetishized tolerance of risk as a part of the performance of the powerful man; risk became part of the astronauts's identities, forming what Tom Wolfe's novel (1979) (and 1983 film adaptation) famously referred to as the 'Right Stuff.' Yet this attitude towards risk was certainly not blind masochism; it was, rather, predicated upon a set of techniques concerned with the control of risk wherein the astronauts were rigorously tested, and trained, to manifest a high degree of corporeal control and calculation over their own bodies and perform tasks in this hostile environment —to maintain control in a situation despite the danger and get a job done. The same, of course, was true of Kratz's mission-control team or von Braun's Saturn V production team; it was not risk *per se* that was being prioritized within NASA but risk management, the search for predictability and control through planning and calculation, whether in price, time, quality or safety. Risk management is at the heart of systems management techniques like PERT and CPM, and this is why Johnson (2002) optimistically suggests: 'Systems management has tamed R&D' (p230-1).

But there is also a distinctive extra-terrestrial masculine mythology at work here: once in Space the astronaut embodies this sense of rational control and emotional detachment. At this point the astronaut has overcome the physical hardship of spaceflight, appears rational, controlled and detached, watching over a chaotic and unstable female Nature below (a 'Mother Earth' —Cosgrove, 2001: 7). With similar conclusions, Baudrillard (1983) recalls the space race: 'we are dumbfounded by the perfection of their planning and technical manipulation, by the immanent wonder of programd development. Fascinated by the maximisation of norms and by the mastery of probability' (p63). Analogous ideas can be found in Barthes (1957) account of the Jet-Man as 'nearer to the robot than to the hero' (p71). His task is to accept and perform 'a kind of vertical disorder, made of contractions, black-outs, terrors and faints; it is no longer a gliding but an inner devastation, an unnatural perturbation, a motionless crisis of bodily consciousness' (Barthes: 1957: 71). For Barthes the Jet-Man is 'defined less by his courage than by his weight, his diet and his habits (temperance, frugality, continence)' as he necessarily conforms to an 'ascetic life' (1957: 71-2). Given the technical and physical likeness of jet testing and space flight, many early astronauts were recruited and tested from jet-test programs (Hersch, 2012: 10-19), including Neil Armstrong who was recruited by NASA from the US Air Force in September 1962. Hersch (2012: 26) suggests that in the 1950s and 60s, American astronauts were deliberately chosen to be courageous, clean-cut, loyal and learned. Hersch (2012: 26) perceives these values as 'characteristically American' (p26), in opposition to a sinister Soviet threat, but we can just as easily describe them as hegemonically masculine, set against a feminine Other.

Within all these images of the astronaut there exist mutually shaping essentializing relations between gender, visuality, Space, risk and technology that prefigure the hyper-masculine, if sometimes contradictory, identities described in the *Right Stuff*; these identities are increasingly evident when they challenged with their Other(s), namely female bodies.[4] Such an instance occurred in 1962. A small group of women, who had previously proved successful in passing

---

4   The bodies of animals, in particular chimpanzees also throw into relief masculinist assumptions surrounding NASA. Indeed chimpanzees were used in the early days of the space program to test the effects of weightlessness for later human astronauts. While the presence and competence of such passive explorers might seem to challenge masculine identities based on heroic action, as Haraway (2004: 93-5) explains, their role is actually more complicit with masculine identities than might be suspected. They were, as she puts it, 'neonates, born of the interface of the dreams about a technisist automaton and masculinist autonomy' (Haraway, 2004: 94). Or, in other words, they were partial subjects whose symbiotic relations with telemetric technology made them appear fully controllable, predictable and ordered, all under the watchful gaze of man (for further discussion of animals and NASA see Gray, 1998).

physical and psychological testing in a privately funded women astronaut study organized by a physiologist called Dr William Lovelace (see Shayler and Moule, 2003; Weitekamp, 2004), sought to join NASA's astronaut program. The Lovelace Women, as they have since been referred to, de-stabilized many of the recurring bodily performances enacted through NASA that prescribed normative gendered assumptions. The desire of these women to become astronauts, to travel in orbit and perhaps even to the Moon, transgressed the tacitly masculine ordering of particular bodies and bodily practices into specific spaces and times under this project of progress and modernity. Instead of the female body being cast to the margins of American progress (to the home or mall, or perhaps, in time, the lunar home or mall), as a reproductive Other, the Lovelace women sought to occupy centre stage in this project of American modernity. These desires can also be, aptly, read as 'lines of flight' (Deleuze and Guattari, 1988: 88-9) that simultaneously transgressed, and exposed, the spatial gendering of banal and messianic space within the image of America as a transcendental state.

In Butler's (1990) terms, these bodies offered hope 'in the possibility of a failure to repeat, a de-formity, or a parodic repetition that exposes the phantasmatic effect of abiding identity as a politically tenuous construction' (p192). Just as some homosexual bodily performances may present a particular body in an opposing gender role (Butler, 1990: 167-70), thus exposing the de-stabilized 'ground' of gender, these astronauts desired to place a female body in a hegemonically masculine project. Yet equally, as Butler (1990) makes clear, such transgressions, while sometimes transformative, are frequently accompanied by 'punishments that attend not agreeing to believe in them' (p190). This was evident in the case of the Lovelace women when in 1962 Lovelace contacted NASA with the chance of taking over the private study. The official reply from Hugh Dryden was 'NASA does not at this time have a requirement for such a program' (quoted in Weitekamp, 2004: 128). Again, the technocratic specter of instrumental rationality is used to conjure up a belief in neutral, automatic and unilateral technical progress as driving forwards the space program. In turn, this meant that the space program could be constructed as if it were an inevitable temporal sequence, expressing natural gender roles and bodily practices, and devoid of ethno-political performativity (Shayler and Moule, 2003: 87).

The disagreement eventually led to a rather absurd Congressional hearing in October 1962, in which the Lovelace Women, led by Geraldine 'Jerrie' Cobb, were examined by Congressmen partly in an attempt to somehow objectively illustrate their technical worth above and beyond their male peers (Shayler and Moule, 2003: 149). Despite demonstrating their capacity to pass flight tests, the women's eventual defeat came in a belief asserted by NASA that astronauts had to be jet test pilots, because only jet test pilots possessed the necessary competencies to undertake high-performance and high-risk flight experiments. At that time women were excluded from becoming jet test pilots because it was deemed too risky (Shayler and Moule, 2003: 149). In this case, the underpinning masculinized relationship between technology and risk proved intractable; accordingly men were able to dictate thresholds of female risk. As Weitekamp's (2004) explains, this process of discrimination was two-fold: on the one hand, NASA seemed reluctant to subject women to degrees of risk because 'the prospect of subjecting a woman to mortal danger betrayed the rigidly defined gendered roles asserted in post-war America' (p3), On the other hand, this paternalist designation of women as needing protecting might itself lead the public to conclude that if women flew in spacecraft then the crafts themselves might be deemed too straightforward and safe. Thus, as Weitekamp (2004) puts it, if 'a woman could perform those tasks [it] would diminish their prestige' (p3). The Mercury astronaut John Glenn, who had just returned back to Earth to a ticker-tape parade after being the first American to orbit the Earth, explained the matter as he saw it:

> I think this gets back to the way our social order is organized really. It is just a fact. The men go off and fight the wars and fly airplanes and come back and help design and build and test them. The fact that women are not in this field is a fact of our social order. It may be undesirable (quoted in Weitkampf, 2004:151).

While Glenn's reference to 'undesirable' may be telling of shifting attitudes towards women, he nevertheless asserts that there is something essentially masculine about these interactions between the body and space technology, so that only particular bodies were deemed not just more desirable but almost factually suitable. Here, hegemonic masculinity was being exposed for all to see, as an necessary ingredient of this figure *par excellence* of American modernity: the astronaut. Even more illustrative, perhaps, of the pervasive, yet slowly changing, masculine/feminine roles across the astronaut program, and indeed NASA, are the accounts of astronaut wives, such as this account by Dotty Duke wife of Apollo astronaut Charlie Duke, recalling her lifestyle during Apollo:

> … to make sure that my husband was taken care of in such a way that he could do the best job possible. I tried not to bother him with mundane burdens at home. Most [astronaut] wives cut the grass, took out the garbage, and kept the house and kids in order. That was our contribution to the U.S. effort in space (quoted in Duke, 1990)

Astronaut wives such as Dotty Duke outwardly accepted their rather stereotypical reproductive, consumer and domestic relations with NASA's astronaut program as almost a social fact, although her reference to cutting grass possibly gestures towards shifting gender roles in this suburban America homes. This acceptance was perhaps fortunate given the lack of seriousness often afforded to the possibility of women, such as Geraldine Cobb, someday working in outer space. Indeed, jokes on the subject were commonplace within NASA, as Wernher von Braun demonstrated in a speech given at Mississippi State College: 'Well, all I can say is that the male astronauts are all for it. And as my best friend Bob Gilruth [director of Johnston Space Center of manned spaceflight] says, we're reserving

110 pounds of payload for recreational equipment' (quoted in Parade Magazine Sunday Supplement, December 1962 —reproduced in Kevles, 2003: 4). Harry Hess, a Princeton Professor and Chair of the Space Studies Board at the National Academy of Sciences, adopted a less jovial approach to explain away female astronauts by stating unequivocally that 'leaving the kids behind was not part of womanhood's idealized image' (quoted in Kevles, 2003: 47).

Ultimately, as Weitekamp (2004) surmised of the Apollo era: 'NASA had no room in its mission objectives for acting as an agent of social change' (p157). Thus, and giving lie to the jet pilot defense, until 1978 'more female monkeys flew in space than female humans' (Hersch, 2012: 152). Indeed it was not until 1978, and the development of the shuttle program, that NASA would select women as astronauts.[5] By this point, and rather predictably in the context of this study, frontier analogies were being drawn upon to retrospectively excuse the omission of women from past astronaut selections: NASA's media rhetoric talked of the shift from explorers to pioneers, or from surveyors to homesteaders (Kevles, 2003: 56). Making a similar nod to spatialized gender roles, Carolyn Huntoon (2002), NASA's first, female, field-centre director (JSC, Houston), describes what she saw as the reasons behind the new policy for astronaut selections to the space shuttle:

> It was going to have more space in it for the crews. It was going to have some of the conveniences of home that previous space capsules had not had. And the laws were changing in our country that women could no longer be discriminated against. The decision was made that we would select qualified women to fly in space (p10).

Again the domestication of space missions appears to go hand in hand with the presence of women in outer space. In both cases, stereotypical gender roles, frequently made through a gendered (mis)reading of American frontier expansion in the nineteenth century, provided an ill-fitting though seemingly seductive temporal analogy to explain away almost thirty years of institutionally prejudicial accounts of bodily difference and space exploration. This re-telling of a highly gendered, spatial division of labor, as a temporal sequence, where male explorers precede female pioneers, reveals the way prescriptive bodily performances were retrospectively enacted across NASA.

Eventually in 1983, over thirty years after the Lovelace women's plight, Sally Ride would become the first American female astronaut to fly into space; though

---

5   During the period before 1978, steps towards eradicating essentialist gender roles in NASA's selection process had been slow. For instance, two 'Lovelace' scientists in the mid 1960s recanted on their early work and testified to NASA that female astronauts were prone to be more 'emotionally unstable,' as supported by 'unspecified studies of menstrual women' (Kevles, 2003: 13). This study is lamented by Kevles (2003: 13), principally because of its tautological argumentation —because there were no female astronauts in NASA then why would further research be required to disprove such prejudicial accounts.

not until Eileen Collins in 1999, over 40 years since the founding of NASA, would NASA give a woman the opportunity of commanding a spacecraft (Kevles, 2003: 13). Kevles (2003) optimistically argues 'women can now decide risks for themselves' (p56). Many more women are now entering science and engineering disciplines and training to be pilots; yet only 10 per cent of astronauts being selected are women (Hutton, 2002: 12); perhaps the instutionalized desire to uphold this assemblage of American modernity as a male domain is changing, if slowly.

Technocracy Challenged

Technocratic thinking helped to establish NASA during the Apollo era as the supposed guarantor of an American transcendental state; yet it also perpetuated a set of narrow assumptions about what individuals this supposedly American led, better, world would value most. These assumptions involved essentializations of Space with masculinity as well as whiteness and Christianity (Hersch, 2012). Of course, this is perhaps hardly surprising given the culturally narrow provenance of American transcendentalism (as described in Chapter 1). Moreover, the masculinization of Space did not simply reflect or represent pre-existing gender relations but performed them in a particularly important manner. Namely it assimilated gendered relations under the messianic ideals of the transcendental-state and so rendered them predestined; yet these bodies were never mere intermediaries of such ideals, they were not internally constituted and fixed in space and time, but relationally constituted mediators capable of modifying these prescriptions, as the desires of the Lovelace women illustrate or the need for the wives of the astronauts to assume male responsibility and continue to take out the garbage or cut the grass until their husbands returned.

During the late 1960s the capacity of technocratic thinking, to perpetuate and obscure discrimination based upon gender, race and class, became increasingly apparent, whether in Space, the streets of American cities or the jungles of Vietnam. A significant element of this resistance occurred in Paris in May 1968, when up to 11 million French students and factory workers, often referred to as the 'New Left,' began a series of demonstrations against various capitalist-technocratic organizations. At times verging on a popular revolution, the demonstrations debilitated France's economy for two weeks, and added to a global tide of political activism against the paternalistic tenor of technocratic instrumental rationality that NASA seemed to epitomize (Feenberg, 1999: 21-43). 1968 illustrates how popular acceptance of the technocratic, big business, big government, mantra of American modernity had begun to unravel (McDougall, 1985: 443-9). The demonstrations in France complemented increasingly violent civil protests in the US during 1968 related to the Vietnam War and civil rights (Bizony, 2006; Burrows, 1998; McDougall, 1985). Symptomatically perhaps, 1968 is also the year that James Webb chose to step-down as NASA's second Administrator after agreeing to make way for a new Administrator when his political ally, President Johnson, decided not to run for re-election, due to public resentment at his handling of the Vietnam War and civil unrest (Bizony, 2006: 212). From the position of many within NASA the tumultuous events of 1968 must have appeared somewhat curious, occurring as they did at a time when the technocratic masterpiece of Project Apollo looked set to cement America's position on humanity's vanguard. In the next chapter I will explain how these changes were not simply passively viewed by NASA but became crucial drivers within the agency for new policies, projects, priorities and values, so as to mitigate political and public criticism. Yet this question remains: how does an organization whose very existence is premised on a heady blend of nationalist mythology and technocracy attempt to modify such potent foundations?

Chapter 6

Was Revolution Ever in the Air?

During the start of the 1970s student political activism against technocracy was prevalent across university campuses in the United States, and was spreading throughout American society. McDougall (1986) explains how anger 'manifested itself in rebellions against traditional curricula, against rationalism, against authority' (p444). Increasingly this anti-technocratic movement spilled over into technophobia, so that 'By the end of the decade [the 1960s] early enthusiasm for nuclear energy and the space program gave way to a technophobic reaction. But it was not so much technology itself as the rising technocracy that provoked public hostility' (Feenberg, 1999: 4). In this chapter I examine how these changes influenced the two-fold messianic-technocratic (de/re-territorializing) relationship between America and Space described in this study up to this point.

Despite the successful Apollo 11 landing, it is difficult to argue that in 1969 the American public, or Congress, would be as receptive to the kind of hyperbolic speeches about Space, as given by President Kennedy in May 1961 (Kennedy, 1961b), where governmental elites readily identified and claimed a space project would secure American values and destiny. By 1969 many of the assumptions contained within Kennedy's (1961b) speech had been challenged by both the space program and events elsewhere. By the end of the 1960s it was increasingly evident that ambitious human space exploration projects would not play a major role in delivering American security or mitigate the Soviet threat of nuclear conflict but that events on Earth such as the Vietnam War or the Cuban missile crisis had and would. And perhaps more importantly, it was now much less apparent that space exploration was a natural destiny or end for humanity, along the lines of the transcendental utopian vision drawn earlier by Tsiolkovsky, Bonestell and von Braun. Earlier critical discussion on space exploration by left-wing commentators like the political theorist, Hannah Arendt (Introduction), and the writer, Kurt Vonnegut (Chapter 3), now appeared to better match the public and political mood. Indeed, as the preceding chapter explains, the technocratic organization demanded by this messianic vision, appeared to threaten rather than liberate or better many people. Thus the paternalistic technocratic ethos and the mythical aspects of the space program became increasingly contested in public and political debate, especially after the 1968 election of President Richard Nixon (Bizony, 2006; Hoff, 1997; Klerkx, 2005). Consequently, the ambitious trajectory of space exploration, initiated by President Kennedy, was becoming understood for what it always was—a highly contingent political choice—a policy that required sustained political will and vast amounts of tax dollars, making it appear at best superfluous, and at worst detrimental, to seemingly more pressing issues of social justice and national security.

In response to this growing impasse between technocracy and democracy, Congress and successive administrations from Nixon onwards (Hoff, 1997), along with the American public (McCurdy, 1997; McDougall, 1985), exhibited rather different attitudes towards American space exploration than Kennedy or Johnson. Increasing apathy towards Space is often cited as part of a wider anti-technocratic, anti-government, counter-culture, movement in America. McDougall (1986) compares the 'stubbornness of the North Vietnamese to the anti-intellectual rebellion on the campus, the public's boredom with space, the perturbing interests and exploitation of the [technocratic] system by the mediocre, lazy and corrupt' (p444).

Public boredom, even hostility, about Space was crucial in enabling the downscaling of NASA's post-Apollo plans to proceed with rather minimal public, and by extension political, opposition. Several commentators (for example Bizony, 2006; Burrows, 1998; Hoff, 1997; McCurdy, 1997) have since observed that while most of the American public remained interested in space into the 1970s, the kind of passionate fervour displayed in the 1950s was missing. For example, a survey on the 6th October, 1969, in *Newsweek* found that 56% of a sample of 1321 readers believed less money should be spent on space, despite the popular Apollo 11 landing less than three months earlier (Newsweek, 1969; for similar opinion polls see Hoff, 1997: 120; Nye, 1994: 242-3). Yet despite such indifference, rather than being dismantled entirely, the technocratic settlement, and messianic, nationalistic aura, which surrounded the space program was re-negotiated by political leaders, and NASA.

Space for Change

Unsurprisingly, in the late 1960s, within NASA itself, utopian reveries for space exploration were rampant. Indeed many senior NASA officials, including von Braun, were formulating even more ambitious plans to further the American lead in human space exploration as a matter of destiny. After the success of Apollo 11 in July 1969, von Braun prophesized that 'By the year 2000, we will undoubtedly have a sizeable operation on the Moon, we will have achieved a manned Mars landing and it's entirely possible we will have flown with men to the outer planets' (quoted in Burrows, 1998: 431). In the spring and summer of 1969, NASA's third administrator, Johnson appointee, Thomas Paine (1968-70), and George Mueller (Associate Administrator of the Office of Manned Space Flight, 1963-69), along with von Braun, produced a twenty-year plan for renewed lunar landing, an orbital space station and human missions to Mars (Klerkx, 2004: 281; for an earlier draft see Paine, 1969: 513-9). This ambitious plan was pitched to a Space Task Group (STG) set up by President Nixon, headed by the Vice-President, Spiro T. Agnew, to formulate post-Apollo space policy on August 4th, 1969—15 days after Neil Armstrong had set foot on the Moon.

On the 15th of September 1969, the STG issued their final report to President Nixon. It proposed a variety of options all of which precluded the kind of investment levels required to enable the Paine-Mueller-von Braun plan, instead it called for NASA to re-direct its focus away from what it termed 'crash programs' of planetary exploration and develop a '... balanced program' (STG, 1969: 523); where options for human exploration, such as a re-usable shuttle vehicle and a modular space station facility, would complement rather than detract from scientific goals, international co-operation and military security program (p524). The findings of the STG were far from unexpected, at least for Nixon: despite Agnew's public support of future Mars landings, the STG had from its inception been instructed to report on potential cost reductions in the space budget (Hoff, 1997: 98). In the short term, the findings of the STG meant that the planned lunar landings of Apollo 18, 19 and 20 would not take place: by January 1970 with the support of Congress, Nixon instructed Paine to plan how to cut 25% from NASA's budget over the next two years (Logsdon et al, 1995: 544). As a result NASA's lunar technology was re-designated to develop a small, scientific, orbital lab called Skylab as well as the Apollo-Soyuz Test Project (ASTP). These smaller-scale missions fulfilled the recommendations of the STG for small-scale, cost-effective projects and, in the case of ASTP, the détente policy towards the Soviet Union pursued by the Nixon administration (see Logsdon et al, 1995: 512-59, and also Logsdon, 1995).

The clamor for more ambitious space exploration within NASA during the late 1960s and into 1970 is reflected by the level of shock and dismay across the space agency when public and political support for the space program appeared to be evaporating just at the height of the agency's technological achievements in the Apollo landings. Paine's disaccord with the Administration's lack of ambition for space exploration affected him to such a degree that he resigned just over a year after Apollo 11, on September 15th 1970, and was eventually replaced by James Fletcher, who was chosen to be more circumspect about NASA's future (Bizony, 2006: 220; Hoff, 1997: 100; Klerkx, 2004: 144; Ward, 2006: 267).

Crafting a Compromise

Symptomatic of this shift in ambition was the greatly reduced amount of spending in the wake of the STG for the planned Space Transportation System (STS, or 'Space Shuttle'), which led to a cheaper to build and only partially re-usable design (Burrows, 1998: 518-9). 'That was one of the greatest mistakes that NASA made,' one senior NASA scientist later recalled (Burrows, 1998: 519). This partially re-usable design was eventually approved by Nixon on 3rd January, 1971. Von Braun himself had actually supported the concept, though only as a brief stepping-stone to a fully reusable craft, and eventually a Mars mission (Ward, 2006: 276). During the Shuttle negotiations between NASA, the White House and the Office of the Management of Budget (OMB), the White House often had to defend NASA against further cuts to its budget as proposed by the OMB that would have led to the scrapping of Apollo 16 and 17, and the STS (Hoff, 1997: 108). Central to the negotiation of a compromise to keep the Shuttle, was the eventual appointment of the new NASA administrator James Fletcher in September 1970, who was seemingly more adept at political conciliation, or career protection (Hoff, 1997), and sought to find new ways of accommodating downscaled political ambitions towards outer space within a vision that continued to develop human space exploration.

The result of these political negotiations towards compromise is exemplified in the design of the most recently used (though retired in 2011), and arguably innovative (Launius, 2006), American human space exploration system: the Shuttle. The Shuttle's eventual form was a long way from the fully reusable design concept first proposed by NASA in 1968. The first proposed fully re-usable design specified two independent fully powered craft, one to 'piggy-back' upon another to orbit; this design concept was undoubtedly more costly to build, and technically challenging, yet likely cheaper to run, capable of more frequent flights, and avoided many of the safety problems later associated with the Shuttle (Burrows, 1998: 518-9). However in the context of Nixon's cuts to NASA's budget, this design would prove impossible for NASA to fund and so was replaced with a proposal for a reusable orbiter with three new main engines, powered by a large, disposable, external liquid-fuel tank, and two 'strap-on' solid fuel rocket boosters (Bizony, 2006: 219-22). This design was, as Burrows (1998) wryly observes 'a compromise that would have far-reaching consequences' (p519). Put more bluntly, Bizony (2006) suggests 'it was a dangerously imperfect piece of technology' (p221).

This trend towards compromise meant that rather than abandon faith in technocracy, NASA re-negotiated a new vision for space-age technocracy, the Shuttle itself being the most visible result of that strategy. This new vision packaged outer space and NASA's technical expertise within a much more commercial and security driven ethos. Capitalistic and militaristic agendas, specifically, the development of the burgeoning telecommunications and defense satellite industries, would enable NASA to fulfil a refined, messianic-technocratic vision of exceptional American prosperity and security. Significant to this vision was the proposal that the Shuttle could be a 'do-it-all' spacecraft and NASA a 'do-it-all' organization; the Shuttle would combine the separate satellite launching and human space exploration activities within the space agency, and would also be capable of launching military satellites for the Department of Defense (Launius, 2006; Woods, 2009). From the start critics argued this design approach was flawed: there was never a convincing technical or economic argument in favor of launching a civil or military satellite with all the equipment required for human space exploration, or indeed vice-versa (Bizony, 2006: 221). The reasons behind the eventual design approach were largely political:

Essentially NASA had to tailor the utility of the Shuttle to cater not only for its own interests but also for the interests of others. It had to ally with those who had long fought a consensus based on the exploitation of space (its marketization and militarization) as opposed to the old consensus of exploration and scientific advancement (Woods, 2009: 40-1).

The Shuttle concept was also premised on increased international cooperation as it would be capable of launching satellites and astronauts from different nations: it promised to position Kennedy Space Center as *the* spaceport in a burgeoning global market. As Hoff (1997: 108) suggests, these arguments curried much favour within Nixon's rather more Eisenhower-like vision of the practical rather than symbolic value of Space.

In attempting to build the commercial and military case for the Shuttle between 1970 and 1972, NASA employees, including James Fletcher and George Mueller, employed creative thinking towards cost-benefit analysis projections for the Shuttle's capacity to fly frequently (said to be exceed expendable launchers like Titan or Delta) and deliver cheaper commercial and military payloads in order to persuade the White House and Congress of the Shuttle's worth. In a report produced for the White House on the 22nd November, 1971, Fletcher explained how the shuttle would fly between '30 to 50 flights per year' (see Logsdon et al, 1995: 558) and would therefore allow 'the United States [to have] a clear space superiority over the rest of the world because of low cost to orbit ... [so that] . ... The rest of the world—the free world at least—would depend on the United States for launch of most of their payloads' (p557, see also Logsdon, 1995).

With the eventual backing of the OMB, Nixon approved the revised STS plan on January 5th 1972. In his statement he inserted NASA's newest space vehicle into a well-worn narrative: '[the Shuttle would] help transform the space frontier of the 1970s into familiar territory, easily accessible for human endeavor in the 1980s and '90s' (quoted in Launius and McCurdy, 1997: 238-9). By blending American global leadership in Space, alongside the technological determinism of technocracy, the revised STS plan is far from a radical downscaling of the mythical edifice of the American transcendental state, despite Nixon's (and later President Ford's) professed pragmatic, anti-technocratic, small-government stance (Bizony, 2006; Hoff, 1997 and Woods, 2009). Indeed some senior ex-NASA employees have celebrated these early forecasts for the Shuttle as little more than mythical speculations necessary to win political support (Burrows, 1998: 520; Woods, 2009: 29-30). Even as early as 1985, Alex Roland, a space historian, described the Shuttle as misguided, uninspired and wasteful (Roland, 1985). Yet in the early 1970s the case for supporting the Shuttle consisted of far more than how well it functioned as a more cost-effective space launcher; it also became involved in boosting the aerospace economy and presidential campaigns. The recession in the aerospace industry mapped onto key marginal states in the presidential election of November 1972 (Hoff, 1997: 109). Thus the construction of the Shuttle was attractive to Nixon to secure jobs and votes; even today accusations of political-economic nepotism surround NASA (Klerkx, 2005; Parker, 2009b). NASA also benefitted from constructing the rationale for its new spacecraft around a space industry whose long-term future was unavoidably uncertain; hence the flexibility of the Shuttle appeared even more attractive (Woods, 2009: 30).

Shifting Culture

While in many ways the birth and development of the Shuttle project failed to substantially shift an American mythology of cosmic aggrandizement, in more practical terms the project, especially its impending operational phase, did help re-organize NASA post-Apollo by challenging some of the tenets of Apollo-era systems management as described in Chapter 4. Shifts in NASA's organizational priorities occurred from the early 1970s as NASA's administrators started to position the shorter-term marketization and militarization of Space as key imperatives within (and sometimes over) its longer-term, larger-scale societal, and symbolic, value. Epitomizing this shift towards pragmatism, on June 30th 1972 Wernher von Braun, who since 1970 had been working in NASA's headquarters nominally leading long-term mission planning, retired from the agency. Echoing Paine, von Braun cited similar frustration with the lack of ambition by Congress and the White House towards the space program (Ward, 2006: 270-81). Given the coupling of technocracy and the romantic vision of space exploration, exemplified by Apollo, any organizational changes in the form of technocracy in NASA would also have consequences for those seeking to mythologize Space.

An especially significant change occurred from the 1970s as NASA contracted more and more technical and managerial work to the private sector. Corporate contractors had always worked with NASA (and its predecessor, NACA), however during Apollo, for example, they had been employed pragmatically to achieve the lunar landings quickly and efficiently; yet NASA retained substantial numbers of in-house technical employees and always closely supervised contractors (Parker, 2009b; McCurdy, 1993). As NASA budgets were steeply cut from 1969, its 'blue collar' technical and craft workers were replaced by a seemingly cheaper and more flexible contracted technical workforce; this diminished the intimate technical expertise that defined NASA's organizational culture during the Apollo-era (McCurdy, 1993). For example, during the development and operation of the Shuttle, large aerospace corporations, such as Boeing and Lockhead Martin, were involved far more extensively in the construction and the design and operation of the Shuttle than during Apollo, especially through the United Space Alliance (USA) contractor partnership. Since the early 1970s the USA has received over $30 billion in Shuttle development contracts (Klerkx, 2005). More tellingly still, between 1978 and 1989 over 90% of NASA's budgetary increases were channeled into its contractors (McCurdy, 1993: 136); thus even when given the choice in the early 1980s, at a time when budgets began to increase again (up to 1991), NASA chose to contact out more technical work rather than use increased budgets to augment its in-house technical capability. McCurdy (1993) explains that the reasons given for this decision were to reduce costs, especially within the 'so-called Reagan Revolution' (p139) where the private sector was aggrandized within a broader neoliberal, pro-market, political ideology.

The commercial ethos of the Shuttle project helped spur on these cultural shifts, as NASA's stated goal of the Shuttle enabling routine, low-cost, low-risk, profitable space flight, encouraged the development of new contracts to fulfil these seemingly standardized tasks, including launch preparation (McCurdy, 1993: 140; Woods, 2009: 38-9). Initially at least, these organizational shifts appeared highly compatible with notions of an American transcendental state. On July 4th (Independence Day) 1982, as the first space shuttle Columbia landed after its fourth test flight at Edward Air Force Based, President Reagan greeted the crew and proclaimed from his rostrum:

> The conquest of new frontiers for the betterment of our homes and families is a crucial part of our national character, something which you so ably represent today. The space program in general and the shuttle program in particular have gone a long way to help our country recapture its spirit of vitality and confidence. The pioneer spirit still flourishes in America. In the future, as in the past, our freedom, independence, and national well-being will be tied to new achievements, new discoveries, and pushing back new frontiers.
>
> The fourth landing of the Columbia is the historical equivalent to the driving of the golden spike which completed the first transcontinental railroad. It marks our entrance into a new era. The test flights are over. The groundwork has been laid. And now we will move forward to capitalize on the tremendous potential offered by the ultimate frontier of space. Beginning with the next flight, the Columbia and her sister ships will be fully operational, ready to provide economical and routine access to space for scientific exploration, commercial ventures, and for tasks related to the national security. (Reagan, 1982)

During Reagan's presidency, narratives of an American transcendental state, reified through metaphors of the 'final frontier,' persisted in offering *the* means to frame American space policy. However, a counter-narrative continued: pre-empting recent critical discussion on spaceflight (as in Dickens and Ormrod, 2007; Macdonald, 2007; Redfield, 2002), left-leaning historians in the 1980s began to debate the merits of this persistent historical analogy:

> ... although in space there are no Indians and no plasmoid buffaloes to exploit, the only nations that can afford to make use of the potential material wealth in space are those that can now afford the enormous expense to reach them. It is likely that in exploiting space we shall continue the same imbalances of resources and material wealth we experience on Earth (Williamson, 1987: 260).

In a rare, and brief, critical discussion of the frontier mentality in outer space, Launius and McCurdy (1997) précis the potent critique of the 'final frontier' metaphor drawing upon the historian Patricia Nelson Limerick: '[the final frontier] denotes conquest of place and people, exploitation without environmental concern, wastefulness, political corruption, executive misbehaviour, shoddy construction, brutal labour relations, and financial inefficiency' (p239-240).

As if rehearsing these sentiments, NASA stumbled and faltered. McCurdy (1993) observes a series of related consequences to NASA's cultural shifts towards a professed 'operational mentality,' exemplified by the Shuttle project, including; reductions in flight testing, a lack of tolerance of risk and failure, tendencies to recruit on the basis of managerial rather than technical skill, decreased organizational flexibility, a lack of central controls, the prevalence of an institutional protectionist attitude in NASA field centres and poor communication. All of these changes challenged the centralized, formal, highly-structured, mission orientated, and classically technocratic, system management approach, which, as Johnson (2002) explains, helped lead NASA to success with Apollo (Bizony, 2006; Parker, 2009a). And yet, while McCurdy (1993) suggests the turn to an operational mentality was problematic, and in the case of the losses of the Challenger and Columbia Shuttles, tragic, it is too simplistic to say that ultra-technocratic, systems management, is a pre-requisite of a successful space organization *per se* (a point also suggested by Parker, 2009a). Recent analysis by McCurdy (2013) has suggested that intensive, less formal, co-located, team-based management (the 'Skunk Works' approach) have been a successful, less costly, and quicker, alternative to Apollo-era systems management, albeit perhaps only on smaller-scale projects, most notably the 1997 Maths Pathfinder project, as well as a wave of recent so-called 'Space 2.0' entrepreneurial ventures like Virgin Galactic (Parker, 2009b—in Bell and Parker, 2009a).

Exemplifying some of the organizational tensions at stake in these shifts in and around NASA is a controversial example of a mid-1990s contracted project to replace the Shuttle. The project, named the Space Launch Initiative, consisted of a competition for aerospace contractors to develop a fully re-usable space plane that would be more efficient (in terms of running, payload costs and flights per year) than the shuttle. After a design competition the project was won by Lockhead Martin in 1996 with a proposal build a space plane called the X-33; however the project was cancelled by NASA in 2001 due to cost over-runs (NASA had spent almost $1b in the project before it was cancelled). Klerkx (2005: 100), and other space commentators (e.g. Stine 1996; Sietzen and Cowing, 2004), suspect that funds were deliberately mismanaged by Lockhead (and even NASA itself) so that the project could not pose a threat to Lockhead Martin's own interests in continuing the potentially more lucrative Shuttle contract, while also preventing other companies outside of the USA partnership from doing similarly. These design competitions were intended to maximize efficiency and ingenuity, and indeed they serve as a precursor for current NASA programs, like Commercial Orbital Transportation Services, that encourage contractor design efforts to develop human space exploration; however in the case of X-33 the project failed to deliver cost savings and technological development, as both NASA and Lockhead appeared to protect their own interests in preserving pre-existing funding streams, let alone fulfil some grandiose vision of America or humanity in Space.

Notwithstanding such venal episodes, as Reagan's hyperbolic comments at Edwards Air Force Base in 1982 demonstrate, the difference between NASA's market-adapted vision to enable American global leadership in the space and telecommunications industry, and Kennedy's 1961 vision of American identity being aggrandized in outer space, is easily overstated. Both visions are highly compatible with a sense of America as exceptionally transcendental: a nation set apart, located on humanity's spatial and temporal frontier. However it is important to recognize that post-Apollo the messianic, mythological overtures of Kennedy-era space policy were far harder to sustain without co-opting the rhetoric, if not reality, of Space marketization and militarization. Equally while belief in the deterministic destiny of technology, a crucial tenet of faith in technocracy (as explained in Chapter 4), was perhaps receding in NASA as budget cuts moved NASA into a more protectionist, less mission orientated, modus operandi (McCurdy, 1993), such technocratic faith was hardly absent within space policy debates: NASA continued to advance nebulous notions that the Shuttle would fundamentally change America and change Space (Woods, 2009). Moreover, deeply held beliefs within successive administrations, from Nixon onwards, in the essential value of a suite of hypothetical, hugely ambitious, anti-ballistic missile space technologies, as part of the Strategic Defence Initiative (later, National Missile Defence), rehearse a technocratic ethos in Space surely just as passionate and free-floating as any advanced by President Johnson or James Webb (compare with Burrows, 1998: 532-43; and Peoples, 2009).

In perhaps one of the boldest recently published incarnations of this messianic-technocratic vision of America in Space, in 2006 NASA's Chief Historian, Roger Launius, produced this homage to the legacy of the Shuttle on its 25th anniversary:

> Because of its technological magnificence, the Space Shuttle has become an overwhelmingly commanding symbol of US technological virtuosity. Ask almost anyone outside the USA what ingredients they believe demonstrate its superpower status in the world, and they will quickly mention the Space Shuttle as a constant reminder of what Americans can accomplish when they set their minds to it (Launius, 2006: 232).

More singularly messianic tropes appear in a recent analysis of the Mars Exploration Rover project, by the NASA cognitive scientist William Clancey. As he reflects upon the potential for robots to replace humans in space missions, Clancey recalls, in quasi-spiritual terms, the indispensable role of the frontier in American culture, hence 'the articulated motivations for human spaceflight will need to shift from talk about scientific exploration to something more deliberately poetic, commercially practical, or frankly political and nationalist' (p249); this is simply because, as he puts it, '*We* are a people who go into space' (p254; emphasis added).

Even within the recent so-called 'Space 2.0' movement (composed of many self-styled libertarian, anti-bureaucratic, anti-government, anti-big business, space entrepreneurs) messianic-technocratic, beliefs in the transcendental power of American space technology, and by extension, and somewhat paradoxically, an American transcendental state, resound (Klerkx, 2005: 298-9; Parker, 2009b: 89-92). Elon Musk, the South African born, billionaire founder of Paypal, and CEO of SpaceX, a company that is the leading constituent of NASA's COTS program, offers his vision: 'The future's going to be really different if we're a spacefaring civilization or not. To me, the idea of being forever confined to Earth is a terrible, sad future. In the future, we should be exploring the stars' (Musk, 2012). Musk's repeated conflation of the human and American 'we' is telling: 'I think that for me and a lot of people, America is a nation of explorers. I'd like to see that we're expanding the frontier and moving things forward. Space is the final frontier and we have to make progress' (Musk, 2012).

And so, while the end of Apollo, the much-cited 'Golden Age' of spaceflight, surely marked a change of emphasis in the way America could be aggrandized through outer space, it hardly represents a decisive break in that political-cultural project. The romantic coupling of exploration and America nation-building appears a too seductive and long-standing narrative to easily abandon or ignore. Even four decades after Apollo evidence continues to suggest that critiquing or simply ignoring this transcendental story appears somehow 'un-American.' In 2010 in a speech at Kennedy Space Centre setting out new plans for American human space exploration in the form of the Orion and the Space Launch System projects, President Obama offered his version of this mythology:

> The space program has always captured an essential part of what it means to be an American -- reaching for new heights, stretching beyond what previously did not seem possible. And so, as President, I believe that space exploration is not a luxury, it's not an afterthought in America's quest for a brighter future -- it is an essential part of that quest ... if we fail to press forward in the pursuit of discovery, we are ceding our future and we are ceding that essential element of the American character.

Our goal is the capacity for people to work and learn and operate and live safely beyond the Earth for extended periods of time, ultimately in ways that are more sustainable and even indefinite. And in fulfilling this task, we will not only extend humanity's reach in space – we will strengthen America's leadership here on Earth. (Obama, 2010)

Despite the shift in space policy and NASA's organizational culture post-Apollo, and its sometimes disastrous consequences (as will be explored in Chapter 8), such sublime, messianic, and thoroughly technocratic, visions, still frame how American political leaders, NASA, and other space proponents, seek to promote the development of space technology to the American public and beyond. But it is perhaps Apollo 11 that will forever be understood as the touchstone of this epic narrative of American identity and destiny. On July 20th 2009, President Obama marked the 40th Anniversary of the Apollo 11 lunar landing by inviting Neil Armstrong, Buzz Aldrin and Michael Collins to the White House. In a short speech, pre-empting his new direction in space policy, Obama proclaimed: 'Apollo 11 is a symbol of what a great nation and a great people can do if we work hard, work together and have strong leaders with vision and determination' (Obama, 2009)

Even though Apollo 11 successfully landed and returned from the Moon over forty years ago, it endures as a cipher for messianic nation building, whether in the speeches of American Presidents, or in the celebrations of technocratic systems management (Johnson, 2002: 230). Yet, the image of America as a transcendental state has always been Janus-faced: it necessarily projects an exceptional, singular narrative linear of American identity and destiny, drawn from selective elements of the past, projected into an uncertain, yet always *better* future. In the next chapter I examine two contemporary sites of memorialization through which exceptional, messianic, cosmic futures, have been produced, as well as some deterritorializing lines of flights that problematize this epic American story. In doing so, the empirical focus of this study will shift away from space policy and NASA's organizational culture and back towards the reproduction of Space within a more public domain of American culture.

Chapter 7
Memorializing the Future

Walter Benjamin's description of Paul Klee's painting, *Angelus Novus* (the Angel of History), can be found within his 1940 essay, *Thesis on the Philosophy of History* (Benjamin, 1970/1999: 249). In this influential essay, Benjamin critiques the progressive orientation of Marxist historical determinism, but his thoughts also support wider analyses of all manner of progressive, linear histories including the messianic mythology of an American transcendental state addressed in this study. Benjamin's poetic Janus figure (Angelus Novus), fixated on the gathering entropy of the past, was powerless to look towards the future. Thus the barbarian past keeps on hurling itself into the present (at the angel's feet): 'There is no document of civilization which is not at the same time a document of barbarism,' proclaims Benjamin (1970b: 248). By describing how a storm is blowing from Paradise, propelling this angel into the future with open wings, Tiedemann (1984) suggests that this allegorical combination of forced flight and an unknowable future (as symbolized by the angel's open-wings), points to a vision of history emphasizing 'super-human despair in the face of the inhumanity of history' where, 'If anything still propels humanity onward, it is the memory of this lost Paradise' (p140). Hence, crucially, this storm of progress renders the future completely unknowable. The angel facing the past, as Tiedemann (1984) writes, 'sees nothing of what is to come' (p141). The messianic teleology of a progressively better future—the core of mythologies of an American transcendental state since Puritanical Divine Providence—is thus rendered impossible.

The resultant effect of the angel of history, for Benjamin (1970b), is to produce "a *weak* messianic power" (p246, original emphasis), contrasted against the strong messianic power of the classical figure of Janus as a gatekeeper between a wretched past and enlightened future, between civilization and barbarism (and of the kind described up until now in previous chapters). Benjamin (1970b) sensitizes us to how arrested, chaotic progressive histories burst into the present, and so rather than time being empty, homogenous—akin to the flight of an arrow—'time is filled by the presence of now' (p253). Hence the historian must 'grasp the constellation which his own era has formed with a definite earlier one' (Benjamin, 1970: 255) no matter how revolutionary or unsettling the results. Tiedemann (1984) sums up Benjamin's critical approach: 'The destructive or critical element in historiography is the exploding of historical continuity' (p157).

Benjamin's (1970b) poetic thoughts orientate this chapter. The mythology of the 'Shining City upon the Hill'—the essence of the American transcendental state described throughout this study—is founded upon the bedrock of a progressive history where America is heralded as the exceptionally better place in humanity's future, whereby cumulative advances in America's space program function as confirmation of that future to come. In this chapter I do not simply ask the extent to which stories of America as a transcendental state have been reproduced in two sites of memorialization—namely the National Air and Space Museum and Kennedy Space Center's Visitor Complex—rather I propose to interrogate how specific practices and spaces in these sites, organize, and disorganize, progressive histories. Following other critical studies of progressive histories in museums (for example Crane, 1997, Hetherington, 2006, Marstine, 2006; Peckham, 2003; Preziosi, 2003; Wallis, 1994; Witcomb, 2003), this task is premised upon Benjamin's critical stance on history: past progress is continually arrested, the future is unknowable, and our civilization is mesmerized by its unforgettable inhumanity.

This critical approach also involves both recognizing that 'museums don't just represent cultural identity, they produce it through framing' (Marstine, 2006: 4) and that in so doing they produce 'particular worlds, histories and peoples' (Preziosi, 2003: 173). Museums can thus provide us with spaces to glorify, celebrate and romanticize 'our' society, and all manner of imperialistic projects undertaken in 'our' name, not least the cultivation of an American transcendental state through space exploration. However, as Hetheringon (2006) reminds us, drawing on Benjamin, we should also bear in mind that even the most imperialist of museums are far from straightforwardly consumed:

> Museums may create a spectacle of experience that is false and imaginary but they produce the effects that can unsettle and challenge these effects just as much as critique from outside can. For example, as the times change, museums often do not, or they change at a slower pace. This can unsettle their own narratives from within and thereby their own conditions of possibility. As their once optimistic narratives of progress and civilization come to be read as narratives of Empire and racism, we see the outmoded speak in unanticipated ways (p601).

Positioned through these theoretical concerns, this chapter consists of two sections of empirical analysis; each will include a brief introduction to the site of memorialization followed by a theoretically informed autoethnography of some elements of that site undertaken between 2005 and 2006. Where possible I have also included some reference to more recent changes at each site; although, as Hetherington (2006) suggests above, museum sites seldom change quickly.

The National Air and Space Museum

The National Air and Space Museum (NASM) building on the National Mall in Washington D.C was opened by President Gerald Ford on July 1st, 1976. It forms the largest museum within the Smithsonian Institution which from the mid-nineteenth century has developed a variety of other major museums in and around the National Mall in Washington D.C., including the National Museum of American History, the National Museum of Natural History and most recently, the National Museum of the American Indian. The opening of the current NASM facility on the Mall postdates the formation of the National Air and Space Museum itself in 1966; however the Mall facility enabled the Smithsonian to bring together a range of flight and spaceflight artefacts that had previously been held at the old Smithsonian Arts and Industries building on the Mall and in a storage facility in Maryland. Across its two sites (on the National Mall and the Stephen Udvar-Hazy Center in nearby Chantilly, Virginia) the NASM holds over 60,000 objects in addition to 12,000 cubic feet of archival information (NASM, 2013). Since 1976, more than 311 million people have visited the museum, a rate of approximately 8 million per year. About 20 per cent of these visitors come from outside the U.S (Romanowski, 2002). In 2012 the NASM ranked as the third most visited museum in the world (Museum Index, 2012). From 2004, these visitor numbers have been augmented by the opening of a NASM annex building, the Stephen F. Udvar-Hazy Center, located approximately 30 km from downtown Washington D.C; although the majority of visits to the NASM still take place on the Mall.

The NASM, along with all other Smithsonian Institution museums, receives its funding directly from the U.S. Congress;[1] as a result, visitors to the NASM are admitted free of charge. All income from sales in the various shops at the museum also supports the chartered educational purposes and activities of the Smithsonian Institution. Although it is principally a museum, the NASM also operates as an internationally recognized research and educational institute, in accordance with the Smithsonian Institution's pedagogical remit. In the case of the NASM, the focus is upon aviation, spaceflight and space-science history.[2] Its academic esteem is furthered by the fact that the museum is also the federally mandated repository for all civilian American space artefacts, important archival documents and many other government mandated flight artefacts.[3] Accordingly, it also actively stores

---

1 Although the operating costs of the Smithsonian Institution are all paid for by Congress, private or corporate financial support or sponsorship is frequently used to aid expansion, such as the Stephen F. Udvar-Hazy Center, named after the Californian business man whose record donation of $60 million to the Smithsonian Institution paid to construct the facility (Romanowski, 2002: 182, for more information on this Center see Allen, 2004). Other donations have taken the form of artefacts. For example in 2005 the Ross Perot Foundation donated an extensive collection of Soviet space program artefacts to the NASM (Romanowski, 2002).

2 For example, the Smithsonian History of Aviation and Spaceflight Series is cooperatively produced between the Aeronautics and Space History Division, who maintain the NASM's archival research programs, and contributing scholars. The series is published by the Smithsonian Institution Press (Romanowski, 2002: 16). Titles include Howard McCurdy's (1997) *Space and the American Imagination*.

3 It is written into Federal law that the Smithsonian NASM has first refusal on all objects no longer required by NASA, from the Apollo capsules to the soon to be retired space shuttle.

and restores space and flight related objects, and loans them out across regional, national and international museum networks. In fact only about 10 per cent of the objects the museum holds can ever be on exhibition at any one time (Romanowski, 2002: 11-5).

Writing in the official museum guide, the director of the museum, John Daily explains the museum's overall mission: 'The museum is here to bring you a first-hand impression of how aviation and space flight have changed the ways in which we travel by air, prepare for national defense, study the Earth and its resources, and explore the solar system' (quoted in Romanowski, 2002: 1)

From the very start, however, the NASM appeared to be about something much more than simply providing an educational catalogue of space achievement and space spin-offs within the transportation, defense and science spheres. On July 1st 1976, two hundred years after Congress passed the first vote for the Declaration of Independence, the ceremonial opening ribbon for the Mall building was cut by a robotic arm donated by NASA from a Viking spacecraft, while on the same day a Viking spacecraft landed successfully on the Martian surface for the first time (Romanowski, 2002: 5). In this highly symbolically charged event, the NASM was being placed as part of an epic story of American progress. In his dedication speech President Ford's spared no opportunity to elaborate this story:

> For three and a half centuries Americans and their ancestors have been explorers and inventors pilgrims and pioneers always searching for something new across the across the continent across the solar system across the frontiers of science beyond the boundaries of the human mind confined within these walls and windows are the products of American men and women whose imagination and determination could not be confined. There is nothing more American than saying: if at first you don't succeed, try, try, again.

> Our bicentennial commemorates the beginning of such a quest. A daring attempt to build a new order in which free people govern themselves and fulfil their individual destinies. But the best of the American adventure lies ahead. Thomas Jefferson said: "I like to dream of the future better than the history of the past" (Ford, 1976)

The museum's institutional mission is perhaps more aptly described as an effort to establish a context for how achievements in Space could be interpreted as part of something broader, as a linear progression of events that address much more fundamental questions than, for example, technical advances in rocket-propulsion technology, questions indeed of who *we* are and what is *our* future. In so doing, the museum, often invites us to read *every* advance in space exploration, as part of a progressive, and indeed often messianic, American story of self-discovery. It is towards these attempts that I now turn.

Remembering Progress

On entering from the National Mall, visitors to the NASM cross into a large, brightly lit, glass and concrete hall (Hall 100) containing a variety of air and space craft (Figure 7.1). These craft are placed either on the ground or suspended from the ceiling as though captured frozen in time mid-flight. While each object is accompanied by a brief textual synopsis of its historical importance, it is the relations between the objects that are crucial to understanding how space and time mutually produce one another in this hall. All visitors to the museum find themselves walking past these artefacts, whilst the other halls—separated into various historical themes such as the space race or planetary science—could easily be missed (Romanowski, 2002). This entrance exhibition is named 'Milestones of Flight.'

Figure 7.1   Milestone to Flight exhibition, Hall 100, National Air and Space Museum, The National Mall, Washington DC, 2005

Source: Photograph by Eric Long, National Air and Space Museum (NASM 2005-176), Smithsonian Institution

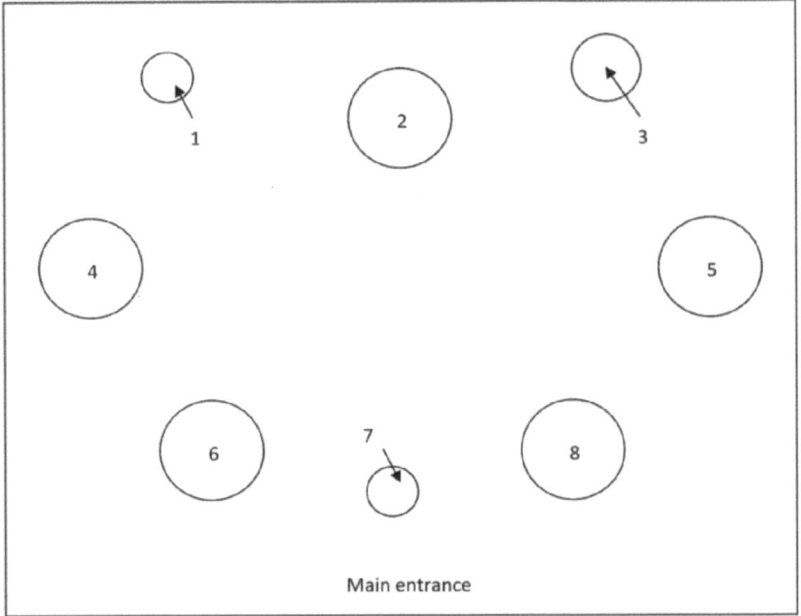

Figure 7.2   Plan of Hall 100, Milestones to Flight, floor exhibits

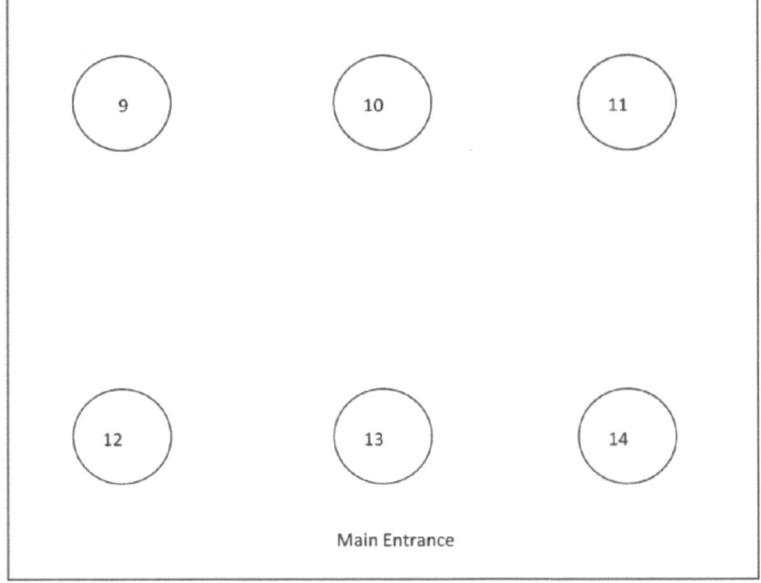

Figure 7.3   Plan of Hall 100, Milestones to Flight, suspended exhibits

Key to artefacts and year of last service use:

1. Two Goddard Rockets (replica, 1926 and original, 1941)
2. Apollo 11 Command Module *Columbia* (1969)
3. Viking Lander (replica, 1976)
4. Pershing II and SS-20 Intermediate Ballistic Missiles (1987)
5. Breitling Orbiter (1993)
6. Gemini 4 Capsule (1965)
7. Touchable Moon rock (from Apollo 17, 1972)
8. Mercury/Friendship 7 Capsule (1962)
9. X-15 Experiment Aircraft (1967)
10. Pioneer 10 (replica, 1983)
11. Spirit of St. Louis (1927)
12. Bell XP-59A (1942)
13. Mariner 2 (replica, 1962)
14. Bell X-1 (1947)

Since the original study was undertaken in 2005, the suspended section of the gallery now also contains the first commercial human spacecraft, SpaceShipOne constructed by Scaled Composited (2004), and replicas of Sputnik 1 (1957) and Explorer 1 (1958).

By placing these exhibits in this central space the curators intend all visitors to the museum to regard these objects as vitally important touchstones to all the other narratives being told therein. To an extent this act corresponds to Witcomb's (2003) insistence on how many traditional museums position the visitor 'as a receiver of knowledge, as the end point of the production process rather than in an interactive relationship to the objects being displayed' (p128). However, while the objects being displayed in Hall 100 are strategically placed as a kind of central pivot of influence to the other narratives in the museum, they are not spatially arranged chronologically into a linear history. But despite the lack of an overtly linear structure to the gallery, no doubt explainable as much by the constraints of the building itself, Hall 100 explicitly rehearses a progressive history. The exhibition display panel in Hall 100 reads: 'These air and space craft are milestones in the history of flight. Each marks an important advance in technology or a historic event. They are a tribute to the imagination, skills and dedication of all men and women who have made flight possible.' Yet this story is not just about the objects but also concerns how we view ourselves, as it continues: 'The history of flight is not simply a story of machines and progress. It also reflects individual ingenuity, personal courage and military requirements and national commitments.'

This framing of the objects in Hall 100 by the associated display panel reveals how the story of spaceflight is being made familiar as an epic story about

American values rather than simply technological advance. Thus McMahon (1981) suggests the NASM 'proclaims the fact and the value of technological progress' (p281), accordingly the museum asks 'young people to aspire to great deeds of technological innovation and thus to keep America great' (p293). This agenda is far from unique. Wallace (1996) explains how the Smithsonian Institution has a long history of dealing with 'technological artefacts—situating them on a progressive continuum from rude to complex, savage to civilized' (p77). Notwithstanding the lesser known British origins of the Smithsonian Institution itself (Smithsonian, 2013), and the recent inclusion of a replica of Sputnik 1 in Hall 100, it is hard to not notice the NASM's nationalistic orientation. The progressive narrative waiting to be discovered in Milestones to Flight, from Goddard's crude rockets that look like little more than glorified fireworks, to bulky, scorched spacecraft like the Apollo 11 Command Module, to the sleek privately-funded spacecraft of SpaceShipOne, offers a version of how Americans can celebrate and discover their exceptional potential by crossing new frontiers in Space; this narrative journey starts with homespun inventors like Robert Goddard, moves to state-sponsored heroes like Neil Armstrong and now onto innovative entrepreneurs like Burt Rutan (the founder of Scaled Composites). This well-worn mythology of the American frontier from early pioneers to state conquest to private settlement (Parker, 2009b: 89-90), invites the audience in Hall 100 to view America as 'a cause, not a nation' (Kaldor, 2003: 12). A 1979 guide to the NASM suggests how this progressive history is almost ritualistically consumed:

> Each person who visits the National Air and Space Museum finds himself moved by that experience in a way he may not have anticipated, affected personally by the sudden, unexpected intimacy of his contact with history—history which, in some cases, is so recent that it is not surprising when a museum visitor is seen reaching hesitantly upward toward a spacecraft's heat shield as if it might still be warm to the touch (Bryan and Dixon, 1979: 20).

In-between the dash for the in-house fast-food restaurant and the gathering of belongings after being processed by the airport style X-ray scanners and metal detectors at the main entrance, Hall 100 does facilitate such sublimely romantic myths and rituals, even when the spacecraft are displayed behind protective Plexiglas. Yet for some the 'tone of celebration and amusement' (McMahon, 1981: 295) on offer in the NASM has proved more problematic.

One particularly strong challenge to the progressive tone of the museum occurred after the announcement in 1995 that the Enola Gay would be put on display in an exhibition about the Atomic Age and origins of the Cold War. Debates were drawn between curatorial staff, including the NASM's then director, journalists, and scholars, who wanted the aircraft to be represented with 'balance,' so as to implicate it fully in the destruction of thousands of lives and the heralding of the Cold War threat of nuclear destruction, and veteran organizations who believed the plane should be simply celebrated as a heroic catalyst for victory, peace and the saving of life at the end of the Second World War, with no, or minimal, mention of Japanese causalities (Gieryn, 1998, and Wallace, 1996: 270-318). The ensuing debate was immediately polarized between those who eulogized the NASM for its patriotic, celebratory and progressive view of American technology and those who lamented this restriction. One supporter of the veterans suggested how the NASM should not exist as a 'countercultural morality pageant put on by academic activists . … who will not give up their radical political agenda' (Gieryn, 1998: 220; and also Crane, 1997).

Eventually, due to the significant political and financial support the veterans associations could muster, including the threat of repercussions to the funding of the NASM, the decision was taken to construct a display of the Enola Gay providing minimalist factual reference to its dropping of the first atomic bomb on Hiroshima, and a justificatory video of its actions by its flight crew. After the controversy, Martin Harwitt, the then director of the NASM, who was forced to cancel his preference for a more 'balanced' display was forced to resign (Gieryn, 1998: 220). For Wallace (1996) this revised exhibit made it appear 'as if the plane that dropped the atomic bomb were an artefact akin to a kettle or a wedding dress, which required only some donor-provided information about its original usage' (p301). The change appeared to confirmed Wallace's (1996) ringing denouncement of technology exhibits across the Smithsonian, as 'sanitized versions of technological history, bereft of human agency, assembled artefacts in evolutionary array,' pronouncing 'progress inevitable' (p82). Nevertheless, the incident did provoke a scholarly, if not public, debate around the relationship between collective memory and interpretive history (Crane, 1997; Prosise, 1998). The Enola Gay is now exhibited, minus the crew video, at the Udvar-Hazy site of the NASM (Allen, 2004). Despite the enduring memories of this controversy, the NASM has since displayed many other less-than-progressive histories. These have, as Crouch (1997) suggests, in admiration of a revised WWI display, intentionally moved away from 'Jewel Box' philosophy by 'confronting illusions and harsh realities' (p6), including those related to spaceflight. In the section I will discuss some of these exhibits as they relate to space technology.

Barbarians in the Hall

Two of the more surprising objects on display in the NASM can be found towering over the left hand side of the Milestones of Flight gallery as you enter from the Mall: the American Pershing II and Soviet SS-20 intermediate-range ballistic missiles (Figure 7.4). These missiles stand in marked contrast to the recognizably romantic and celebratory narratives of exploration and bravery contained in the hall. They are painted in camouflage and as such they bear little resemblance to the nearby Mercury, Gemini and Apollo spacecraft. These missiles might easily be expected to challenge the celebration of technology as both emotionally and morally progressive, because, as with the Enola Gay, their presence here is haunted

Figure 7.4   Pershing II and Soviet SS-20 Intermediate Range Ballistic Missiles in Hall 100, National Air and Space Museum, Smithsonian Institution, 2005

Source: Photograph by author

by the on-going threat of a nuclear apocalypse that they enabled. However, such emotional anxieties are immediately mitigated: when we move closer to the stand we can read a display panel informing us that these particular missiles were decommissioned as part of the 1987 Cold War Intermediate-range Nuclear Force Treaty (INF). And so, as the official guide puts it, they stand in this room as 'a milestone in the effort by the United States and Soviet Union to control nuclear arms' (Romanowski, 2002: 30). In an instant the question of nuclear war appears foreclosed and America's exceptional moral and emotional destiny is rehabilitated.

Witcomb (2003: 138) suggests how strong linear narratives are regularly organized in museums in the manner of the two missiles on display: shared emotional identifications are offered to direct us to moral answers so as to inhibit critical interpretive debate and conjure up the self-affirmation of an audience's values. And yet the missiles enduring presence in the second decade of twenty-first century, almost thirty years after the 1987 INF, just as easily invites very different thoughts: the continued threat of nuclear destruction, global nuclear proliferation,

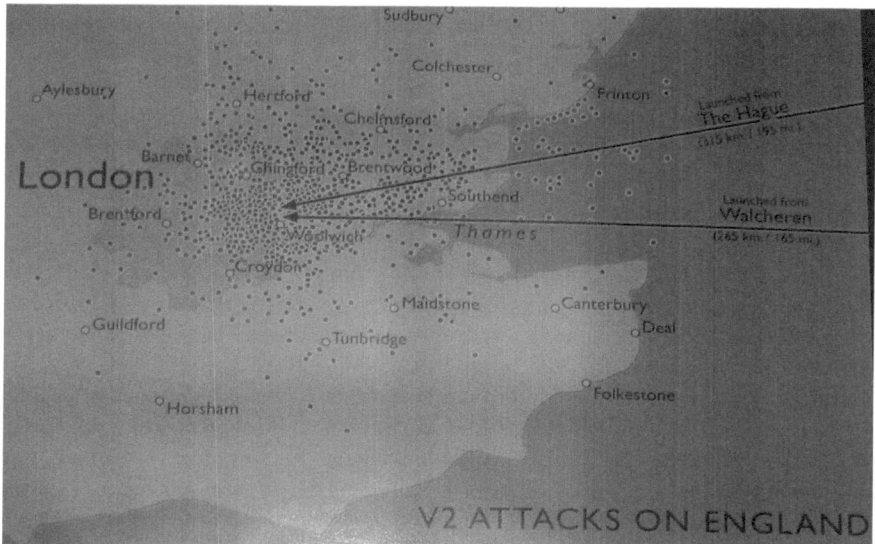

Figure 7.5   V-2 exhibit in Hall 114, National Air and Space Museum, Smithsonian Institution, 2005

Source: Photograph by author

debates around national missile defense, the weakness of recent arms reductions treaties, and the cost of preserving a nuclear deterrent and decommissioning.

Some similarly uneasy objects are found in Hall 114 which since the museums beginnings has hosted a large exhibition entitled 'Space Race,' containing objects and textual displays intended to provide a historical narrative of the development of space flight immediately after the Second World War. These narratives center upon key NASA programs: Apollo, the Shuttle and the Hubble Space Telescope. In this hall the origins of these civilian space projects are linked to the rivalry between the two superpowers; as a result the ostensibly heroic, scientific and peaceful civilian space program is faced with its shadowy older sibling: the militaristic project of designing rocketry as an efficient means of delivering or threatening mass murder. And yet, even here, the NASM attempts to sustain a narrative of moral progress. In keeping with the linear chronologies employed throughout the other halls in the museum, the historical narrative begins at the entrance to Hall 114 with an exhibition titled 'Military Origins of the Space Race' which recollects the development and operation of the V-2 rocket in Nazi Germany, during the final years of the Second World War by von Braun. This display contains a series of information panels beneath a V-2 rocket rebuilt by the US Army.

These panels convey the story of the V-2 in text accompanied with photos (including—since 1990—a dead body from a V-2 strike and concentration camp laborers) and a map depicting the course and impact of V-2 attacks on England

(Figure 7.5). The exhibit contrasts the pioneering, technical development of the V-2 as the first, accurately guided, high-altitude rocket, and its value to American space exploration, against its tragic cost: the use of slave labor, in the development of the V-2 as well as the thousands of Allied civilian fatalities it caused (Cadbury, 2005).

One display panel explains: 'Late in World War II, Germany launched almost 3000 V-2s against England, France and Belgium. After the war, the United States and the Soviet Union used captured V-2s as a basis for developing their own large rockets.' Contrasted against another: 'Concentration camp labourers built V-2s under unbearably harsh working conditions. Thousands perished in the process.' The official guidebook straddles this fraught relationship between the morally abhorrent and the technically progressive, suggesting that the V-2 was 'An ominous harbinger of a new kind of warfare, the V-2 sparked large-scale rocketry efforts in both the United States and the Soviet Union' (Romanowski, 2002: 99). And yet, while the monstrous beginnings of the American space program are acknowledged, no information is provided of the strategies after the Second World War that erased von Braun and his colleagues' support of Nazism and war-crimes and re-activated them at the heart of the nascent American rocket program (Burrows, 1998: 122). Rather a display simply notes that:

> The US Army brought captured V-2 missile parts to White Sands Proving Ground, New Mexico, for its Project Hermes missile development program, managed by General Electric. Wernher von Braun and his team were housed in nearby Fort Bliss, Texas. They advised General Electric personnel in the reassembly, testing, firing, and evaluation of the missiles. The first firings, in 1946, used all-German components. Later, modified American-made components were substituted to gain practical experience and to improve the basic missile design.

By focusing on the period directly after the Second World War, von Braun is depicted here simply as an incidental technical consultant accompanying the rocket parts, instead of the leading technical director of the team that would ultimately launch the first American satellite and develop American rocketry for almost thirty years (Ward, 2006). The text also carefully re-configures the rocket as American, seemingly purging its distasteful past as the barbarian history of another nation by emphasizing how it was upgraded, technologically and morally, with American labor and parts.

The only other reference to von Braun in Hall 114 can be found in an exhibit on the far side of hall from the main entrance where the depiction of von Braun is noticeably less morally ambiguous and the centrality of his work to the American space program can be safely acknowledged. Here von Braun's face appears, taken from a 1958 Time magazine cover (Figure 7.6), in an exhibit comparing his fame and popularity to his publically undisclosed Soviet rocket scientist rival, Sergei Korolёv. The display panel explains how von Braun became director of Marshall Space Flight Center, and development the 'Moon Rocket'—the Saturn V; it

Figure 7.6   Wernher von Braun exhibit, Hall 114, National Air and Space Museum, Smithsonian Institution, 2005

Source: Photograph by author

continues: 'von Braun was an avid proponent of space exploration, in the 1950's he collaborated on a series of magazine articles and television shows depicting future space travel.'

When mentioned in relation to the V-2, von Braun is as a faceless German scientist, once his achievements for the American space program are established we are allowed to see his face, his smile, his determined gaze, and celebrate not just his technical prowess but his desire to popularize American spaceflight on television shows and in popular magazines. By spatially separating out these displays the NASM allows the museum's audience to separate out the barbarian Nazi/German past from the civilized, messianic American future. In other words, a progressive history is resumed; thus allowing us to make 'sense of the relationship between time and space in a modern world where they are thrown continually into doubt' (Hetherington (2006: 601). The appearance of von Braun in Hall 114, and the spatial and temporal (and moral) void that appears between the two displays, suggests how space and time are being organized as a progressive sequence.

After reconciling the monstrous history of von Braun, on the wall on near side of the hall we can find a series of photographic panels from the Hubble Space Telescope (HST). This exhibit complements the HST structural test vehicle which is displayed in this part of the hall. The display consists of several large frames that contain the widely known, photographic 'spacescapes' produced from spectrographic data from the HST. Close to the first frame is a panel that contains an introductory overview of the HST's mission, which seeks to establish the importance of the photographs:

> Launched in 1990, the Hubble Space Telescope was the first optical imaging telescope in space devoted entirely to astronomy. From its orbital path above the atmosphere, Hubble can see more sharply and detect a wider range of colors (from near infra-red through visible light and into ultraviolet) than telescopes on the ground. Hubble has contributed to breakthroughs in many areas of astronomy and astrophysics, including cosmology, planet formation, stellar evolution and black holes. But the telescope is perhaps most famous for its breath taking, close-up images of the universe. On this wall are some of Hubble's best.

These images fatally interrupt or arrest a progressive history; the Hubble photographs span light years, and consist of images that are millions of years old, how do we reconcile them with a mythology of an American transcendental state as conjured up through Moran's depictions of the American West or even television images of the Apollo landings, a mere 250,000 miles away. Perhaps here we are confronted with something much closer to the positive sublime described by Lyotard in relation to modern astrophysics, where human thought is delimited, rather than aggrandized (in a Kantian sense), by technoscience (Lyotard, 1998; for similar motifs see Jorgensen, 2009). Yet reconciliation is viable. We can as Parks (2005) suggests, allow Hubble to return us to the 'flesh or the familiar' (p149). First, the culprit, 'Arresting Images' (Figure 7.7), is identified within the

Figure 7.7   Hubble Space Telescope exhibit, Hall 114, National Air and
**   Space Museum, Smithsonian Institution, 2005**

Source: Photograph by author

exhibition; then the escape is pursued through a visual metaphor involving one
particular Hubble image—the Eagle Nebula.

The accompanying panel reads:

> After the Space Telescope Science Institute released this spectacular portrait of
> the Eagle Nebula in 1995, both NASA and astronomers were surprised at the

public's enthusiastic response. The event became a milestone in the history of public outreach at the Institute. Soon a group of Hubble astronomers and imaging experts banded together to form the Hubble Heritage Team. Since October 1995, they have released new and wonderful views of the universe each month.

Objects imaged by Hubble are often impossibly vast and amorphous—as unearthly as can be. Yet the compositions captured by the Hubble Heritage Team and others tend to evoke landscapes. The press release for the Eagle Nebula compared the gaseous columns to 'towering buttes and spires in the deserts of the American southwest.' Other Hubble images have been likened to roughhewn mounts of gas and dust. This allusion to landscape art connects the Hubble images to the romantic frontier art and artists of the 19th century who celebrate the majesty of the American West

By superimposing motifs of the American West onto the origins of the Universe, the origins of all human life,[4] the American transcendental state not only resurfaces but is conveyed in one of its most powerful guises. Far from being lost in Space, America appears again at the origins of the Universe. Paralleling Lyotard (1998), Parks (2005) speculates that such humanizing acts may be a reaction against the de-centring of humanity by the mediated gaze of technoscience:

Hubble 'Can see both further and better than the human eye,' the satellite threatens to undermine the authority of the scientist-viewer whose knowledge is derived through I. Put another way, since the Hubble image implies the astronomer's obsolesce, humanism must be constantly invoked within it as a way of affirming the human/subject viewer sitting on Earth at this satellite-computer-telescope's other end (p150).

Moreover, as we are invited to reflect on the prowess of American space technology to aid this humanistic self-discovery, we are presented again with a temporal sequence of progress, led by America, which now runs from the origins of the Universe (Parks, 2005: 155).

Moving upstairs beyond Hall 114 visitors find another gallery, Hall 210, which contains an exhibition called 'Apollo to the Moon.' While this gallery contains various pieces of space hardware, including a Skylab Command Module and an F-1 engine from the first stage of the Apollo rocket, some of the most interesting artefacts are those related to the Apollo astronauts, including a test version of the Lunar Rover Vehicle (as used by Apollo 15, 16 and 17). This gallery offers two challenges to the progressive history of spaceflight, both related to the figure of the astronaut. Amongst various items, the gallery contains an exhibit that recreates an

---

4   As with all Nebulas, the concretion of plasmas and dust in the Eagle Nebula enables star formation which in turn enables all the heavier elements that form planets and life to exist.

Figure 7.8   Apollo to the Moon exhibition, Hall 210, National Air and Space Museum, Smithsonian Institution, 2005

Source: Photograph by author

Apollo landing site, containing a spacesuit, an American flag, and other astronaut tools and equipment (Figure 7.8).

This display also includes rock sample return containers from Apollo 12, navigation aids, spacesuits still darkened by the lunar soil, and the Lunar Rover Vehicle, as well as a variety of smaller objects including scissors, sunglasses, documents and used food packages. The nearby display panel explains why Project Apollo flew astronauts: 'The six Apollo lunar landing missions demonstrated the value of manned exploration of planetary surfaces. The astronauts were able to set up scientific instruments, choose the most interesting samples for collection, and study the geology of the lunar surface.'

This explanation for sending people into Space complements a similar display panel at the Udvar-Hazy Space Center which asks the question—'Why Send People?' The panel answers: 'Robotic explorers are less expensive, but they lack the experience and judgment that makes humans valuable in space. The intelligence, skill, problem-solving ability and personality of astronauts have been central to the image and popular support of space exploration.' Or put more directly: 'By promoting the astronaut corps, agency advocates and media leaders were able to reduce complex technical issues to personal values such as bravery and patriotism' (McCurdy, 1997: 88). Yet this display also begs the question of what kind of heroism is on display here.

Hersch (2012: 1-4) explains that astronauts, including those of Apollo, were not recruited because they were mavericks but because they were skilled professional managers, capable of effectively navigating rules, orders and hierarchies within large bureaucracies. The heroism of the astronauts, their autonomy, was a product of their location within a complex socio-technical machine. As Hersch (2009) makes clear, the objects on display here, whether scissors, spacesuits, checklists or food containers, enabled their unique agency, as much as any inherent skill. We can thus understand the astronauts, via Haraway's (1991) insistence that 'The machine is not an *it* to be animated, worshipped, and dominated. The machine is us, our processes, an aspect of our embodiment' (p180; original emphasis). However, within a progressive history such notions appear barbaric, or monstrous: how can Americans contemplate a messianic mythology of progress if their fate is wedded to the mundane vagaries of machines. Thus the unique agency of astronauts must be defined in opposition to the machines that constitute it, as in the display panels quoted above; only then can the mythology of progress continue.

Another troubling exhibit in Hall 210 is the manifestation of the deaths of Apollo 1 astronauts—Ed White, Virgil 'Gus' Grissom and Roger Chaffee—on a Apollo timeline display. While it might be suspected that the loss of these astronauts (and those of Challenger and Columbia since) would challenge the notion of spaceflight as inherently progressive, revealing our frailty and dissolution, instead these deaths appear vital to that noble story—the display notes how the sacrifice of astronauts evidences and supports technical progress: 'The tragedy caused the Apollo program to be delayed many months. The interior design of the Apollo command module was extensively modified and all flammable parts were replaced with flame-resistant and flame-retardant materials.' A nearby panel quotes Grissom: 'If we die, we want people to accept it. We are in a risky business and we hope that if anything happens to us it will not delay the program. The conquest of space is worth the risk of life.'

These statements mirror a display panel found next to two plaques in Hall 113 (the 'Moving Beyond Earth' exhibition), containing mission patches and flags recovered from the wreckage on the final flights of the Challenger and Columbia Shuttles: 'Both times, the commission or board investigating the accident determined the technical and human factors that led to the accident and recommended remedies to eliminate the problems. Each accident prompted changes in hardware design, procedures and management to improve safety.'

Kristeva (1982: 110) proposes that such noble memorials[5] must not allow the corpse, or objects connected to it, to become unrestrained objects of worship;

---

5 Although not discussed in this chapter, Kennedy Space Center visitor complex contains the Astronaut Memorial Mirror. This designated National Memorial represents the largest memorial to astronauts killed in pursuit of spaceflight and contains similar sacrificial motifs, for example on the dedication marble. Other memorials include those at Arlington

rather the artefacts of death (for example pictures of the deceased[6]) must remain taboo, or abject. This is because in any act of sacrifice, the promise of immanence with the power of God ('the One'—Kristeva 1982: 111) can only occur if we counterbalance any thoughts that the sacrificed corpse is 'desirable, fascinating, and sacred' (Kristeva 1982: 110) through acts of making it abominable: 'the abominate is a response to the sacred, its exhaustion, its ending' (p111). The dead astronauts thus operate as a spectre within a progressive, messianic history, occupying 'the borders of our existence, threatening the apparently settled unity of the subject with disruption and possible dissolution' (Grosz, 1989: 71).

What connects all the artefacts and exhibits discussed in this section is the apprehension and mitigation of open-ended, barbarian and heterogeneous time, or in Benjamin's (1970b) terminology 'moments of danger' (p247), whether the on-going threat of nuclear annihilation, von Braun's Nazi past, the inhuman de-centering of the HST, or the hybrid or dead astronaut. Various narrative devices are employed in these exhibits to locate the story of American spaceflight on the vanguard of the morally (and technologically) progressive vision of history found throughout the NASM, these include the use of moral rejoinders, narrative absences and spatial separations, the superimposition of nationalistic mythologies, and myths of heroism and sacrifice. However by bringing these devices to attention in this way I remain uncertain as to the extent that visitors to the NASM unreflectively participate in their reproduction (Hetherington, 2006). After all, contemporary audiences need only consult their smartphones and in a few minutes they can find out exactly how many nuclear missiles are still deployed in silos in the US and Russia, that von Braun's Nazi SS membership was erased by the American government after his immigration to the US (Cadbury, 2005), that the Hubble's imagery may have been deliberately produced from the raw data to appear like those of the American West (Kessler, 2012), and the Apollo astronauts' daily work was far from exciting or daring (Hersch, 2012). In the next section I examine how artefacts at Kennedy Space Center's visitor complex further contribute to the dis/organization of a progressive American history through spaceflight, and by extension an American transcendental state.

Kennedy Space Center Visitor Complex

Next to the NASM, the visitor complex at John F. Kennedy Space Center (KSC), Merritt Island, Florida, stands as perhaps the most well-known public site through which to memorialize, and indeed encounter, American spaceflight. In contrast to the NASM's political sponsorship through the Smithsonian Institution, the

---

National Cemetery and Hemphill, Texas which was the principal location for Columbia recovery efforts.

6   While Hall 114 contains photos of the corpses of the victims of V2 rocket attacks it remains very unlikely that photos of the corpses of astronauts will be placed on public display.

inception of the visitor complex at KSC was rather more ad-hoc. The original plan had its origins in the early 1960s when families of those working at KSC and Cape Canaveral Air Force Station were allowed to drive around the restricted grounds on Sundays (KSC, 2006d). As interest in the space program grew during the 1960s, the American public increasingly wanted to visit KSC, which was and remains the primary base for NASA launch operations, and in the case of Cape Canaveral Air Force Station, commercial and military satellite launches.[7] In response NASA established a temporary visitor complex nearby and began bus tours of KSC on July 22nd 1966. This complex was then made permanent on August 1st 1967 (KSC, 2006b; KSC, 2006d) and was moved and expanded again in 1971 in response to the nearby opening of Disney World (the latter being the world's most visited tourist attraction). KSC is not unique in having a visitor complex, in that almost all NASA field centers have some form of visitor facility, yet the complex at KSC is the most visited and largest, at 70 acres and attracting up to two-million visitors annually (KSC, 2005a: 8). About 30 per cent of these visitors are from outside the United States, which is the highest percentage of international visitors to a single site in the extensive Florida tourist attractions market. The top three countries of origin in order of numbers are the United Kingdom, Canada and Germany. Two other notable differences to the Florida-attractions market are a higher than average visitor age and that the decision to visit KSC is predominately made by men rather than women. The latter point is, of course, again suggestive of the masculinization of the space program and high technology, as noted in Chapter 5.

The administrative background of the complex differs somewhat from the NASM in that it has never been operated by a publicly-funded pedagogic organization, but instead has been operated by a series of profit-orientated hospitality, catering and entertainment companies and corporations, contracted by NASA. Since 1995, the complex has been run by Delaware North Companies (DNC).[8] While in the past

---

7   It is important to distinguish between the Air Force Station, named after the promontory barrier island on which it is located, called Cape Canaveral, and NASA's Kennedy Space Center. The latter is located a few kilometers north west of Cape Canaveral, across the Banana River on Merritt Island covering 34,000 hectares of land, swamp and waterway. The former was established out of Cape Canaveral Air Force Station by the Department of Defense during the 1950s and KSC in 1962-3 to meet the burgeoning demands of using the massive Saturn V rocket for the final phase of the Apollo program. Since its creation, Cape Canaveral has been responsible for launching all NASA human space flights prior to Apollo 8, as well as most non-human NASA flights, commercial and military launches using expendable rockets, such as Delta, Titan and Atlas. In contrast, KSC has been responsible for the launch, final assemble and final astronaut preparation for all NASA human space-flight operations since the first Saturn V flight of Apollo 8. KSC also acted as the primary processing, launch and landing facility for NASA's Space Shuttle and will be employed similarly for all future Ares-Orion launches (KSC 2006a; Burrows, 1998; KSC 2006b KSC, 2006d).

8   Delaware North Companies is one of the world's largest hospitality and food-service providers. The company is entirely privately held, with an annual revenue of over

contractors have received some public funding and offered free admission to the complex to visitors, DNC has made the complex entirely self-funding through ticketing, food purchases, event bookings and merchandising (see KSC, 2006d). Importantly however, despite contracting out the visitor complex to DNC, NASA representatives still have the ultimate authority to decide on the curatorial content of the complex.

Above and beyond the contractor relationship with NASA, another notable difference to the NASM is that the complex seeks to offer a broader set of experiences to the public to remember the space program. While the NASM visitor experience focuses almost exclusively on exhibits and education, through both static and interactive displays of space hardware from space-suits to space-craft,[9] the visitor complex at KSC offers similar exhibits alongside an array of other experiences that blend education and entertainment. For example, the complex offers: tours of launch sites and working NASA facilities, an astronaut hall of fame, luncheon events with real astronauts, photo-opportunities with astronaut impersonators, a live theatre production for children, full conference and catering facilities, a Shuttle launch simulator, a space camp for children, an open-air memorial to deceased astronauts, as well as the chance to watch NASA launches in person (also KSC 2006c; KSC 2005a). Despite the fact that the complex does not enjoy the constitutionally mandated access to NASA artefacts like the Smithsonian Institution, it still encourages educational links through a NASA Education Resource Center which is located at the visitor complex, and involves NASA educational staff (KSC, 2006c).

During an interview for this project, both Daniel Gruembaum, the head curator at KSCVC and Debbie Land, the chief marketing officer for KSCVC, specified the mission statement as telling 'the NASA story.' Yet in marked contrast to the NASM's pedagogical focus both emphasized the need for more than just story-telling: 'Do we have to border on entertainment? Definitely. Definitely. We have to have a high amount of quality entertainment just to attract the audience. Because you can tell a great story but if no-body's listening. Yeah you've got to have the audience' (Gruembaum, Chief Curator KSCVC) 'We are not a museum exactly, we're not a theme park by any means but we are an attraction. And our goal is everyday how do you tell the NASA story and be true to that story and be authentic while at the same time do it in an entertaining way' (Land, Marketing Manager, KSCVC).

As a private company in the midst of a highly completive Florida tourist attractions market, directly influenced by NASA's current activities, as opposed to a well-established, publically funded museum on the National Mall, the curatorial

---

$2 billion dollars and over 40,000 employees in the United States, Canada, the United Kingdom, Australia and New Zealand (KSC, 2006c, see also KSC, 2006e)

9   One notable exception at the NASM is the IMAX theatre at the downtown location. The visitor complex at KSC also contains an IMAX theatre, frequently showing the same films as the NASM.

and management team operates a site in a very different context to the NASM. The unique blending of the commercial pressures of the complex alongside the space activity at nearby KSC, and the rather ad-hoc origins of the site all engender a rather different visitor experience than the NASM.

Rockets and Rides

When driving along NASA Parkway to the visitor complex, after perhaps catching a distant view of the immense VAB from the bridge crossing the Indian River, one of the first 'Space' related sights is the towering appearance of a series of rockets above the trees prior to the entrance to the visitor complex. These rockets are part of a Rocket Garden attraction which has been present at the complex since its permanent founding in 1967. The Rocket Garden appears as a kind of open-air mausoleum for rocketry, a place where rockets long-since required by the space program are put on open air display for visitors. The Rocket Garden is a fairly unique space for the KSCVC (and NASM) in that it contains very little in the way of narrative structure either in terms of a set route for visitors or display captions. As such, visitors are free to walk around a variety of rockets and spacecraft. What is perhaps most striking about the Rocket Garden is not so much the lack of an overarching narrative structure, but the thoughts that are enabled in its absence.

The rockets found in this garden might be expected to function as symbolic steps in the remembrance of space flight, operating as part of a technological and moral journey towards American progress, akin to the NASM's Milestones to Flight exhibition. But instead, free of an organizing epic narrative we are able to reflect upon a history of progress that necessarily involves obsolescence, decay and death. As I wandered quietly around these former totems of progress, I witnessed their slow decay in the Florida sun (Figure 7.9). These rockets perhaps speak more to Benjamin's (1970a) critique of progressive history where the angel of history is mesmerized by the rubble of past arrested progress gathering at her feet in the present, but unable to repair that progress as she is propelled onwards. Composed of mostly expendable launch vehicles designed to spectacularly burn-up in the Earth's upper atmosphere, the objects in the Rocket Garden were never destined for gentle retirement like NASA's fleet of Space Shuttles or the Apollo Command Modules. Hence we can ask: what happened to these rockets; why were they not flown and expended like countless more; why was the great expense and technical ingenuity involved in their construction deemed unnecessary; what was their forgotten promise. These questions remain unanswered and hence the Rocket Garden arrests the progressive history of spaceflight rehearsed elsewhere in the NASM and KSCVC with a vision of a future that is already obsolete.

Most visitors to KSCVC will experience the Rocket Garden, if only fleetingly as they walk to other attractions. Other aspects of the site require more sustained engagement, not least the KSC tour. Since its creation KSCVC has offered a bus tour of the adjacent NASA launch operations and processing facilities at KSC,

Figure 7.9   Rocket Garden, Kennedy Space Center visitor complex, Florida, 2006

Source: Photograph by author

initially these were run by TWA who also provided some facility management at KSC (KSC, 2014). The current tour is described enthusiastically by the marketing team as 'A must-see experience, this tour takes guests on a narrated video supplemented bus tour of Kennedy Space Center' (KSC, 2005a). Similar promises pervade the official guidebook:

> Kennedy Space Center is the last stop on Earth for space travellers before their journey to the stars. It is here that mission hardware, experiments, astronauts and satellites are brought together ... A tour of the Space Center will show where these dedicated people work and reveal many unusual buildings—some that look odd, some massive and some mysterious (KSC, 2006a).

The tour-guide bus offers a narrated commentary with an on-bus video guide to the sights, and three possible stop off points.

As well as offering technical information, the narrated commentary frames the sensory and emotional experience of visitors. A prominent feature is the hyperbolic descriptions of the various NASA facilities. Continual emphasis is placed on the immense size and quantities, including: the 3.6m m³ Vehicle Assembly Building

**Figure 7.10   Vehicle Assembly Building as viewed from tour bus, Kennedy
              Space Center, Florida, 2006**

Source: Photograph by author

(VAB) used to assemble both the Apollo-Saturn V and Shuttle spacecraft prior to launch (Figure 7.10), which we are told is large enough to produce internal clouds that produce rain; the 300,000 gallons of water released seconds prior to launch at launch complex 39 to suppress damaging sound waves; and the 5,000,000kg of spacecraft and mobile launch platform moved at 1 mph by the squat crawler-transporters between the VAB and the launch pads. Gruembaum explains why physical size figures prominently in the emotional codification of the visitor experience: 'people need to get an emotional connection ... at the visitor complex it's the size of the rockets, the presentation.'

Told in this manner, the 'NASA story' is portrayed as not simply being about reaching towards the unknown, the story is in fact un-representable, sublime, perhaps only partially accessible to our understanding, better suited to our emotions; but in telling this story we can marvel at the reason which constructed such sublime machines. On the KSC tour bus, visitors can collectively consume (again) the American technological sublime, as described by Nye (1994: 237-52) with reference to the launches of Apollo 11 and the Shuttle. Yet, while Nye (1994: 246) is surely correct that the sublime power of a launch, even an unsuccessful one (p251), rehearses the Kantian dynamic sublime in all its sensual terror, the bus tour perhaps approaches something more akin to Kant's mathematical sublime, told in the dizzying magnitude of immense volumes, weights, speeds, and rocket thrust as expressed by the tour narration. In both cases it is American technology and American engineers that appear sublime, capable of producing 'an infinite and perfect world' (Nye, 1994: 287).

As the tour continues, visitors are able to stop at one of the newer visitor sites at KSC, the Apollo-Saturn V Center, constructed in 1996. This building is located adjacent to the now retired Shuttle Landing Facility and is only accessible via the bus tour. Like the bus tour itself the Center invites visitors to consume the American technological sublime. After leaving the bus, visitors are channelled into a hall and are shown a film offering an overview of the history of American human spaceflight up to the early Apollo flights. Visitors are then instructed into a second presentation in a gallery made to resemble the Apollo Launch Control building, complete with re-furbished control Apollo-era consoles and monitors, as well as three large monitors depicting a Saturn V launch, narrated by Apollo 8 (and 13) astronaut, Jim Lovell. This film ends in the deep, booming reverberations of the launch, and perhaps something approximating the dynamic sublime of the launch of Apollo 11 (Nye, 1994: 246).

The tour then proceeds into a large brightly lit plaza which is dominated by a restored Saturn V rocket suspended on steel supports above the floor, placing it just out of reach. We are informed that before 1996 this rocket was displayed next to the VAB, while the decision to house it inside this building was no doubt partly to enhance its preservation, the effect of locating such a large rocket in a building which can only just accommodate it, also usefully enhances its size (Figure 7.11). The guide book explains this attraction thus: 'The centerpiece of this spectacular facility is a fully restored Saturn V moon rocket ... visitors enter the Saturn V

Figure 7.11   Apollo-Saturn V Center, Kennedy Space Center, Florida, 2006

Source: Photograph by author

rocket plaza where they can walk around and under the enormous moon rocket …
Giant rocket for a giant leap—these five State One engines generated 7.5 million
pounds of thrust at liftoff' (KSC, 2006b).

Despite the evocation of the American technological sublime, in this
exhibition, like the Rocket Garden, this Saturn V rocket's continued existence
provokes questions which remain unanswered: why was this rocket not required
by NASA; was it part of Nixon's downscaling of Apollo; if so, why did this occur?
A great deal of post-Apollo history (as told by Burrows, 1998; Hoff, 1997) is
missing from both NASM and KSCVC. Instead of learning about these more
pragmatic moments in the history of American spaceflight, visitors are encouraged
to celebrate more uplifting episodes across a variety of Apollo related displays
and timelines, including one in which astronaut Ed White (who was killed in the
Apollo 1 fire in 1967), described the feeling when carrying out the first American
spacewalk (see Figure 7.12).

Gruembaum is explicit about the need to connect to the visiting public through
such emotional, and patriotic, celebrations, as he describes his belief in offering
'better understanding, not only of the physical universe, but ourselves.'

The emotional and experiential focus of KSCVC is articulated strongly in a 10-
year Thematic Development Plan (TDP), released in 2005. This plan was created

**Figure 7.12   Apollo-Saturn V Center timeline exhibit, Kennedy Space
Center, Florida, 2006**

Source: Photograph by author

by both DNC and NASA. The plan is designed to 'transform the Visitor Complex
into a place where all of NASA's activities around the world, and throughout the
universe, can be experienced,' creating 'a state-of-the-art interpretative center like
no other NASA visitor center, science museum or attraction in the world' (KSC,
2005b). Beyond the marketing hyperbole, and leaving aside the pragmatic need to
compete with nearby theme parks, the TDP blurs the lines between museum and
theme park where simulated experience meets historical storytelling and authentic
objects (Bryman, 2004; Huyssen, 1995). The CEO, Daniel LeBlanc, explains how
KSCVC 'will be the best place in the world to see, *feel* and understand space ...
The final outcome of the 10-year Thematic Development Plan will be the ability
to tell the NASA story better than ever before' (KSC, 2005b; emphasis added). A
2005 marketing brief explains similarly:

> Inspired by real NASA initiatives and fuelled by the human spirit, Kennedy
> Space Center Visitor Complex is charting a new course to bring far-reaching
> NASA exploration events and triumphs to visitors up-close and personal. The
> journey takes NASA's space exploration story to the next level through new
> technology and interactive experiences (KSC, 2005b).

One upshot of this plan was the construction of the Shuttle Launch Experience (SLE), which opened in summer 2007. The SLE, which is located in main complex site on NASA parkway, was part of a deliberate attempt by DNC to compete with the nearby Orlando theme parks. As Gruembaum explains:

> In order to compete with the market today, whether it's movies, whether it's the theme parks, whether it's just in-house entertainment, you've got to be pretty sophisticated, and people expect a wow factor. People expect to be more than just told things they've got to get some kind of emotional connection.

The SLE comprises a series of simulation technologies whereby 'visitors will experience the real-life, hair-raising movements of a launch' (KSC, 2005b). The marketing team at KSCVC describe how SLE offers an:

> ... authentic simulation [that] will take guests on an incredible and emotional journey of discovery to experience what only a few hundred astronauts have dared: launching into orbit aboard the Space Shuttle. "Crew members" will board the Space Shuttle and strap-in for launch in a one-of-a-kind motion simulator that replicates the sights, sounds, 'G' forces and rattle and roll of lift-off designed to bring the mission alive. After this experience, visitors will never view a launch in the same way (KSC, 2005b).

Another marketing release explains how:

> The Shuttle Launch Experience will create an immersive, first person experience infused with interviews and presentations by NASA Space Shuttle Astronauts, who will share their riveting personal stories with visitors. Using state-of-the-art technology, the result will be a thrilling launch simulation that impacts the body, mind and soul of visitors, and forever gives them a sense of connection to the Space Shuttle program ... This monument to the Space Shuttle program will bring the dream of lift-off to reality ... A team composed of astronauts, NASA experts and renowned attraction designers conceived this authentic simulation attraction like no other in the world (KSC, 2005c).

Gruembaum similarly describes how:

> we want to create an experience that people would feel what it feels like to launch in a shuttle ... now if we were a theme park we could probably say we are going to launch you to Mars but this is a simulation ... this is what it feels like when astronaut's launch.

Land further explains that the SLE is a simulator not a ride because:

if you go to Disney and they put you on a space simulator you're going to learn how to fight aliens ... ours is very different in its story telling. It's designed by astronauts; you're going to learn how the space shuttle operates before you experience what it feels like to launch . ... So the entertainment in the simulator is different because it is authentic. It was actually designed by astronauts.

Whether described as an authentic simulator or historical ride, by providing visitors with a sensory experience approaching the awe and terror of a Shuttle launch, the SLE just as readily functions to celebrate the American technological sublime as it competes with Disney World. If the coldly mathematical sublime of magnitude on offer in the bus tour of the VAB, or the Apollo-Saturn V Center, or in the more recently opened Shuttle Atlantis exhibit, is not enough to convince you of the sublime greatness of American space technology, then perhaps you need to hear, feel and see the terror of this technology for yourself in the dynamic sublime on offer in the SLE. Or maybe, like me, you are left feeling slightly out of place here. As with the NASM, many visitors to these sites are undoubtedly also inclined to reflect upon, rather than identify with (Hetherington, 2006), the nationalistic-sublime atmosphere on offer, especially perhaps the large numbers of international visitors.

All this said, I cannot deny my own feelings of excitement, if not terror, while seeing the scorched Apollo 11 Command Module for the first time or peering at the launch pads from which the Apollo 11 voyage began; however these sites also conjured up other responses in me, including strangeness and boredom. No matter how close I was to Launch Pad 39-A or the Apollo 11 Command Module, I felt far removed from the experiences of Armstrong, Aldrin and Collins, while I was all too aware of how much monotonous preparation was involved in their astronaut training and in the development and processing of their spacecraft. The figure of the astronaut neatly conveys the affective paradox of American spaceflight on display across the NASM and KSCVC, which as Frederic Jameson posits is found somewhere between 'adventure and boredom' (quoted in Parker, 2009a: 325). Although often depicted as the 'all-American hero' embodying a blend of faith, courage and honesty (Hersch, 2012: 26; McCurdy, 1997: 88), the astronaut is simultaneously rendered exciting and strange—a figure that can perform impressive feats, a human part in a technocratic machine whose occasionally sublime experiences lie outside our (and perhaps their) comprehension (Parker, 2009a: 330). Equally, the work of astronauts appears both terrifying and dull: much of their training and flight operations consisting of the perfection of often quite repetitive tasks, punctuated by stress, but always undergirded by the threat of danger, both during spaceflight and testing (Hersch, 2012: 66-68). In the next chapter I examine these traumatic affects more directly by focussing on what are usually regarded as two of the most catastrophic moments in the history of American spaceflight: the losses of Apollo 1 and Challenger. In so doing, I consider how these events ordered and reordered the progressive, messianic, history of American spaceflight, re/organizing the American transcendental state.

Chapter 8
Traumatizing Spaceflight

This chapter is concerned with the traumatization of the messianic mythologizing and technocratic organizing of American spaceflight, by the losses of Apollo 1 in 1967 and the Space Shuttle Challenger in 1986. Both of these events induced a milieu of affective responses in and around NASA. Messianic hope and technocratic confidence around American spaceflight was punctured with shock, trauma and anxiety, fuelling a sense that the results of our action are indeterminate, unknowable, and incapable; that a barbarian future of violence, suffering and despair might supplant the promise of nationalistic progress. But these responses were quickly channelled within heroic and sacrificial eulogizing into a project of individual *and* national redemption. This twin-fold process where, described through Deleuze and Guattari's philosophy, negative affects are released and then captured, in recognizable emotions and thoughts, and put to work augmenting power of the State, is the concern of this chapter.

Several studies exist that diagnose, and promise to resolve, technological and organizational failures within NASA (for example Cabbage and Harwood, 2004; Cook, 2007; McConnell, 1987; Vaughan, 1996). However, these studies are not concerned with how these events were rendered emotionally intelligible as part of the cultural and political framing of American spaceflight. This is an important question to consider because, as Latour (2005: 245) suggests, even a transcendental state has to be maintained; this is because it is only as strong as the most fragile link through which it circulates: 'every launch vehicle ... carried the extra weight of the American dream going awry' (Vaughan, 1996: 388). In this chapter I ask here what happens when one of these links fails; and then, how is the romantic edifice of ideas, emotions, machines and people that surrounds American spaceflight resurrected. After all, as previous chapters have shown, from Bonestell's *Collier's* artwork to the on-going exhibitions in the National Air and Space Museum, and despite later losses, notably the Shuttle Columbia in 2003, spaceflight continues to endure as a touchstone for hopeful reveries of American sublime exceptionalism and technocratic confidence.

Drawing on Deleuze and Guattari's philosophy, I explore how the shock, trauma and anxiety associated with the losses of the crews and hardware of Apollo 1 and the space shuttle Challenger, disrupted and augmented hope in America as a transcendental space and time. These feelings are understood here as relational affects, rather than individualized, owned, emotions. Deleuze and Guattari suggest that all recognizable emotions arrive from these affects, but our emotions cannot contain them (Massumi, 2002: 35). This Deleuzoguattarian concept of affect is thus drawn upon here to register the social organization of relational, pre-cognitive, and experimental forces that prime bodies to act or think, without narrowing analysis to recognizable emotions possessed by human subjects (Anderson, 2006; Connolly, 2002; Massumi, 2002; Thrift, 2008). Moreover, these forces are not solely mediated through human bodies: 'thought is bound up with things: it is through things that we think. Then, things act back' (Thrift, 2008: 117). This emergent 'acting back,' and its unpredictable influence upon intelligible cognition and recognizable emotion, is the realm of affect. Hence my analysis goes beyond following the ebb and flow of individual emotional responses in and around NASA to the losses of Apollo 1 and Challenger, to also consider how the emergent, future-orientated, 'eventfulness' of these two events was mediated through various humans and non-humans that primed us to respond in some manner with recognizable thoughts and emotions (Thrift, 2008: 116).

The Apollo 1 fire, January 27th 1967

On Friday, January 27th, 1967, three astronauts, Gus Grissom, Ed White and Roger Chaffee, were strapped into their Apollo 1 space vehicle for a series of tests on launch pad 34 at Cape Canaveral Air Force Station. The astronauts had been chosen to be the first to fly the newly constructed Apollo Command Module; they would also likely be the first choice crew for the planned lunar landing (Burrows, 1998). The January 27th test was a so-called 'plugs out' test; this involved separating the vehicle from the launch tower systems, so that it could operate on internal power, and, once the hatch was sealed, use the same high-pressure, pure-oxygen, atmosphere mixture as flight. Except for some minor communications problems, the launch tests proceeded as expected. The astronauts' bodies were kept regulated in an optimal symbiosis with the machine in which they were seated through various systems designed to monitor their bodily and cognitive capacities to perform during launch. Burrows (1998) describes the purpose of the test as 'learning its [Apollo's] peculiar idiosyncrasies and the location and use of the hundreds of switches that controlled its eighty-eight subsystems' (p408). However in this particular test, unbeknown to the astronauts, their bodies had been plugged into a machine that was about to violently disorganize.

At approximately 6:31pm a fire engulfed the capsule. NASA ground crew standing outside reported bright yellow flames inside. It eventually took just over four minutes to open the capsule due to the pressure difference with the outside, and only then could NASA confirm that all three astronauts had died. A post-mortem revealed that the cause of death was asphyxiation by the toxic gasses released in the fire; the astronauts also suffered extensive second and third-degree burns (Young, 2002; Hines 1967, Burrows, 1998). The visceral trauma of the event threatened to diminish the rising messianic hope and technocratic confidence within and around NASA. The trauma of Apollo 1 threatened to spread with speed beyond NASA, disorganizing, and de-territorializing, assumptions of confidence in the agency to render the future, the cosmos, as knowable, as progressive, as hopeful, as American. NASA aimed to slow this disorganizing affective force through censorship.

In the minutes and hours after the Apollo 1 fire, NASA censored the release of information about the accident to the news media. The American Society of Newspaper Editors, for example, commented how 'Although the agency knew within five minutes, it took two hours to learn all three astronauts were dead. And this came round about from Houston Manned Spaceflight Center rather than Cape Kennedy' (quoted in Burrows, 1998: 409). Through this process, NASA sought to expunge the visceral details, the negative 'affect', of the event. To elaborate further, affect is defined here through Deleuze and Guattari's (1988: 256-7) philosophy as a set of relational, pre-cognitive, and experimental, forces which influence how bodies, human and non-human, can realize *potentials* beyond their actual states and known capabilities to affect one another and be affected (Thrift, 2008: 177-181). Understood in this way, affects were essential to the Apollo 1 test; it functioned as a disciplinary test, intended to confirm that the bodies of the astronauts could become something beyond their known terrestrial capacities: productive parts of the complex machine in which they were placed. Here the 'body's capacities to affect and be affected are entrained through a series of repeated, cyclical steps' (Anderson, 2012: 31), intended to augment some capacities (for example concentration on procedures, calculation, esprit de corps within the team) and diminish others (for example fear, boredom). Through this test the astronaut's bodies were also plugged into the wider body of the transcendental state—they were being summoned to become Gods. Thus we can identify two general types of affects, positive and negative. Deleuze and Guattari explain, affects always have a tenor: they can be positive, joyous, hopefully, 'augmenting' our potential to act, or negative, painful, sad, 'diminishing' that potential (Deleuze and Guattari, 1988: 256). In their words:

> We know nothing about a body until we know what it can do, in other words, what its affects are, how they can or cannot enter into composition with other affects, with the affects of another body, either to destroy that body or to be destroyed by it, either to exchange actions and passions with it or join with it in composing a more powerful body (Deleuze and Guattari, 1988: 257).

The body of the American transcendental state, as described across preceding chapters, involves the production of positive, open-ended affect—conceptualized here as a 'messianic hope'—wherein astronauts bodies, and by extension those of all Americans, might become divine, infinitely capable to act and think, to transcended space and time, through space technology. Anderson (2006: 749) suggests positive effects such as hope frequently accompany imperialist geopolitical projects, especially when they promise a better future elsewhere in time and space, un-related to negative affects from which they emerge and depend (that is, suffering and injustice). The Apollo 1 test released negative affect—traumatic suffering—against, but also within, a space of messianic hope:

**Figure 8.1   View of Apollo 1 vehicle after the fire, 28th January 1967,
            NASA Image: GPN-2000-001834**

Source: NASA

a hopeful machine painfully reduced the astronauts (and itself) to cold, inactive, silent, immobile, matter (Figure 8.1).

In an attempt to block, or inhibit, this transmission of this negative affect, to maintain the hopeful transcendental state, NASA released reports that the astronauts had died instantly in the fire. And yet eyewitnesses would later report seeing the arms of the astronauts flailing around inside the capsule as the fire burned (Anon, 1967b); evidence from inside the vehicle showed they had clawed for up to 20 seconds at the hatch kept shut by the pressure required to maintain the oxygen-rich mixture inside the capsule (Maxwell, 1967). The astronauts' final words to launch control were also sanitized by NASA: at first they were reported as 'fire in the spacecraft' while later reports revised them to: 'We're on fire. Get us out of here" (Maxwell, 1967; Anon, 1967b). A *Chicago Tribune* editorial published a few days after the fire attempted to explain NASA's reaction: 'when information is tightly controlled, as by NASA, it suggests a conscious effort to conceal unpleasant facts or at least withhold them until they can be fitted together in the manner most favourable to NASA and the Moon program' (Maxwell, 1967: 4). Burrows (1998: 409) is less suspicious—NASA's failure to tell the truth was simply because its employees were stunned into lack of action, unable to react, unable to make sense of what had happened. But even four decades later the bodies of astronauts, and their final moments of suffering, disappear from most analysis of this event, and other space-related traumas.[1]

In part we can understand the removal of the visceral details of the death of the astronauts with reference to Kristeva's (1982) notion of abjection, wherein noble, nationalistic, sacrifice requires the corpse must not become an object of worship and fascination (as described in the previous chapter). But Deleuze and Guattari's notion of affect suggests something else was at stake here: by denying their prolonged, and witnessed, visceral suffering, the trauma of the dying astronauts appeared like an instantaneous, localized, mechanical failure; seemingly un -diminishing the potential of anyone or anything else to act or think. And so the troubling negative affects that might circulate beyond the vehicle, affects which seemingly ran counter to the messianic hope organized through the space program, could at least be slowed until they could be captured and put to work. But despite NASA's efforts in this regard, negative effects were released across the news media. A special issue of *Life* magazine, published on the 3rd February, reported how NASA's technology could just as equally create a 'murderous furnace' as instigate our future in space (Maxwell, 1967).

While the press became partly gripped by the visceral suffering of the event itself; politicians translated these negative affects into recognizable heroic emotions and feelings, in order to augment messianic hope. This process began with the sacrificial valorization of the dead astronauts as national heroes (Hines, 1967; Maxwell, 1967). President Johnson paid rousing homage to the astronauts in the day after the fire, telling reporters: 'Three valiant young men have given their lives in the nation's service. We mourn this great loss and our hearts go out to their families' (Hines, 1967). The front-page of *The Washington Evening Star* on January 28th 1967 played on similar heroic sentiments, this time using Grissom's own words: 'If we die, we want people to accept it. We are in a risky business, and we *hope* that if anything happens to us it will not delay the program. The conquest of space is worth the risk of life' (Hines, 1967; emphasis added).

NASA then initiated an investigation into the event, the Apollo 204 review board, which sought to determine why the accident had occurred. The analysis of causes was vital to re-establish a progressive history of spaceflight, technocratic confidence in NASA, messianic hope in America. This is because it relegates the diminishment of our potential to act (negative affect) to a mysterious past; rather than allow the past to continually burst forth into the future draining the future of its positive potential for our confidence and hope. NASA's management thus sought to identify the cause of the fire as quickly as possible through its own expertise. In an effort not just to improve on the design for later missions, but to counter the

---

1   Dickson (2001: 219-20) for example describes, the astronauts dying in seconds or instantly despite the presence of contemporaneous eyewitness testimony and evidence to the contrary (Maxwell, 1967; Anon 1967b).

notion that NASA, and America's, future was diminished. However, unfortunately for NASA, this investigation was far from conclusive as Maloney, a reporter for the *Washington Post*, explains: 'though the investigation was considered the most thorough of any accident, the board could not be specific about the cause of the fire in the cabin of the Apollo command module' (Maloney 1977; see also, Anon 1971b).

NASA's investigation described a complex web of influences, in which various managerial breakdowns culminated in an array of technical oversights, namely that uncertified, highly flammable materials made their way into an oxygen-rich atmosphere inside the capsule (see NASA, 1967). In response to the investigation in NASA, the House of Representatives and the Senate began parallel investigations seeking to explain the incongruous situation in which a group of world class scientists and engineers, funded by the wealthiest nation on Earth, overlooked a startlingly simply problem: 'electrical wiring in an oxygen atmosphere where flammable materials is present constitutes a dangerous situation" (Sehlstedt, 1969).[2] These investigations did not sit easily alongside the emotionally uplifting eulogizing of the crew within memorial ceremonies, as they asked whether the noble mission that the astronauts had died for, was itself the cause of the fire. A *New York Times* article published just days after the fire suggested that two Apollo engineers believed the pressure of the space race had fostered compromises over safety (Wilford, 1967). Later, in 1971, both *Life* magazine and the *Los Angeles Times* revealed how an ex-NASA subcontractor named John Dietz, repeatedly spoke up about the safety concerns of using highly combustible materials in the oxygen rich capsule; and yet NASA's organizational hierarchy deemed his concerns of minor importance to the technical progress of the program as a whole (quoted in Anon, 1971a; Anon, 1971b).[3] Moreover, Dietz was reported in *Life* magazine as stating that even if his warnings had been heard, 'Everybody was so bound to a schedule that I don't think they'd have stopped the test' (quoted in Anon, 1971b). The underlying criticism of NASA was that technocratic confidence, that is faith in scientific and technological expertise as *the* solution for societal ills, was being implicitly diminished, not augmented (see Chapter 4), by the urgency, at least partly, induced by the messianic hope (against latent Cold War anxiety) of Kennedy's 'moon shot.' On the 10th anniversary of the fire, the *Washington Post* asked explicitly: 'DID THE HASTE in trying to achieve the Kennedy-set goal of having men go to the moon by 1970 contribute to the accident? Did trying to beat the Russians to the moon cause the deadly mishap?' (Maloney, 1977; original emphasis).

---

2   The Congressional investigation was broadly critical of both NASA and its contractors. For example, Representative Teague (D., Texas) described the Apollo program as replete with 'All kinds of carelessness' (Anon, 1967a).

3   The centerpiece of the *Life* story was a letter from Dietz to NASA which focused on the problems of using an insulating material (U-577-1) on circuitry in the capsule which was ignored by his line managers (Anon 1971b).

Elsewhere the confidence of technocratic organization, not messianic hope, was the focus of criticism. During the Apollo 1 investigations Congress reproached NASA for suppressing a report in 1965 by the Apollo organizational consultant, Samuel Phillips, that criticized the performance and progress of North American Aviation—the manufacturers of the Apollo capsule (O'Toole, 1967a).[4] Moreover, the fact that this report was suppressed from Congress (see Reistrup, 1967b) contributed to a feeling that NASA was becoming more and more institutionally aloof, operating with the pretence of an omnipotent technocracy that the Senate report characterized as 'overconfidence and complacency' (Congress Report, 1968). During the late 1960s, various Congressmen cited NASA's arrogance over Apollo 1, in particular that of James Webb, as responsible for the reduction in NASA's budget in 1968: 'Instead of sympathizing with him over the fire (which they did at the outset of the hearings) they turned on him because of what some of them called 'his obsession with secrecy' (O'Toole, 1967b: A1). Such feelings surfaced again within a series of civil lawsuits for corporate negligence filed against North American Aviation (NAA—the manufacture of the Apollo 1 capsule) by the families of the three astronauts (Anon 1971c). Within these lawsuits John Dietz provided 'an unsettling eyewitness account of carelessness and bureaucratic indifference that invited disaster' (Anon, 1971b). For Burrows (1998) the negative affect of Apollo 1 fundamentally weakened American technocratic confidence in technology: 'The deaths of Grissom, Chaffee, and White showed that technology that was supposed to serve its creators by carrying them to the stars could just as easily and without warning become a dangerous, malevolent, unpredictable monster that defied orders with terrible results' (Burrows, 1998: 412).

Yet, as McCurdy (1997) suggests, in many ways 'Rather than reduce support for the civilian space effort … the tragedy actually served to strengthen public resolve, as opinion polls revealed' (p102). But the Apollo 1 fire was not without its wider consequences. Organizationally, the space race was slowed down significantly: NASA eventually spent over $700 million in 19 months after the Apollo 1 fire to modify the Apollo spacecraft (Anon, 1968) and did not launch a manned Apollo flight until almost two years after the Apollo 1 fire (Apollo 7). Apollo 1 also heightened the pressure of NASA to control its work more fully, leading to more intensive mechanisms to produce calculable space and time within the organization, through the progressive refinement of systems management as Apollo advanced (Johnson, 2002). But the intention to produce these ordered, striated, spaces was now gathered from a milieu of negative and positive effects, the former with instantaneous potential for disorganization, for the potential de-territorialization of the American transcendental state. As William Hines,

---

4   James Webb (NASA administrator) told a senate committee that the 'Phillips' report was acted upon by NASA and indeed a Senate committee revealed this to be the case, while Webb explained its secrecy as vital to maintain relations with contractors. However, Congress reports would later connect this attitude towards secrecy as damaging the responsibility towards safety within NASA (O'Toole, 1967a; Reistrup, 1967a).

reporter for the *Washington Post* warned after the success of the next manned Apollo mission (Apollo 7)—NASA must never forget the 'frightful memory of the dying screams of the men on Pad 34' (O'Toole, 1967a: A9). Hope now emerged through not against, or despite, suffering and trauma (Anderson, 2006: 749); thus perhaps hope was rendered increasingly more messianic through NASA, even as confidence within NASA appeared less assured.

Challenger, January 28th 1986

Just under nineteen years after the Apollo 1 fire, on the 8th January 1986, the Space Shuttle *Challenger* (STS[5]-51-L) made its way to launch pad 36B. The launch of the 25th Shuttle flight was set for 22nd January. On-board Challenger were seven crew members, including six full-time astronauts and Christa McAuliffe, a 37 year-old-high-school teacher. McAuliffe was to be the first participant in the 'Teacher in Space' program, announced by President Reagan two-years earlier during his successful election campaign. The program was conceived to re-stimulate popular attention around what was increasingly becoming viewed as a routinely unexciting space program: 'The public had been lulled into thinking the missions no more risky than an airplane flight' (Ad Astra, 1991: 13). Yet others were more sceptical, as Penley (1997) explains: 'The Teacher in Space program—strongly opposed by U.S. educational leaders—was a Reagan-Bush-NASA media circus' wherein McAuliffe 'was selected for her representative mediocrity [she had little scientific or engineering training or interest] and *knew it*' (p23-4). McAuliffe, Penley (1997) argues, was being used to domesticate sublime and grandiose, high-technology within traditional (conservative) American values: the family, motherhood, small town America. But this attempt seemed to be working. Veteran space reporter, William Broad, reported how 'The highlight of the mission is Mrs McAuliffe, whose presence has drawn crowds of tourists, teachers and reporters' (Broad, 1986: 16), not least because the launch was to be beamed live to millions of school children in classrooms across the United States, while 11,000 teachers across the country had been applied to the Teacher in Space program (Penley, 1997: 41).

NASA had set itself the task to 'perform-or else' (Mackenzie, 2001: 145) as it 'cultivated popular interest in particular missions' (Mackenzie, 2001: 149), missions designed to bolster public and political interest in the space program. What was at stake in this performance was not just confidence in the competence of NASA, but rather confidence in American technocracy. Space commentator Malcolm McConnell (1987) offers us one hyperbolically nationalistic (and equally technocratic) account of this vision:

---

5   STS is the designation used for all Shuttle flights, it refers to Space Transportation System. Due to revisions in NASA flight designations, the flight number does not correspond to the number of shuttle flights.

The space shuttle had become the symbol of America's technological and political renaissance, a bold, successful gamble, the quintessence of this optimistic decade. The orbiter rose beyond its own awesome fires, clean, powerful, high above the ashes of military defeat and political turmoil that had scarred the past twenty years. To millions of people, the space shuttle was a patriotic icon, the tangible symbol of what was best in American civilization, a product of daring scientific prowess, free-enterprise innovation, and insightful political leadership. The space shuttle had carried us back to the frontier and made us proud again (p8).

NASA's confident show was already in doubt when the planned January 22nd launch date came and passed. In fact, the shuttle was eventually to sit motionless on the pad for a further six days prior to launch, due to a range of technical and environmental factors.[6] And then on January 27th, the launch was re-scheduled for the penultimate time to 9.38am EST on January 28th. That night the ambient air temperature dropped below 20°F; it was expected to climb just above freezing the next day. Such freezing temperatures are unusual for central Florida, even in January. As the weather chilled some sub-contracted engineers expressed a variety of technical concerns about the safety of launching in such abnormally cold temperatures. Consequently, on the morning of the 28th three ice/frost-inspection teams examined the shuttle's flight readiness in these conditions. A poll at 9.00am by NASA's Mission Management Team finally approved the launch. At approximately 9.00am the shuttle's crew were strapped in and then, at 11.38am, the 25th space shuttle flight launched from Kennedy Space Center towards low Earth orbit (for more detailed information on the events prior to launch see Burrows, 1998; McConnell, 1987; Vaughan, 1996; Launius and Ulrich, 1998). In every split second of flight, millions of parts, unthinkable to any one person in their dynamic relationships, simply had to perform together (Mackenzie, 2001).

For some engineers working at NASA's sub-contractor—Morton Thiokol—the possibility of the shuttle not performing together was far from a remote risk, it was almost demonstrable. Among these doubters were a number of engineers, working for the Utah based contractor, Thiokol, who designed and constructed the solid rocket boosters (SRB's)[7] (McKenzie, 2001; Vaughan, 1996). Through a series of heated teleconferences with NASA on the eve of launch, the probability

---

6 Many of these reasons are discussed in Vaughan (1996: 52) as well as Burrows (1998: 554). They include the delayed launch of the preceding shuttle flight Columbia (STS-61-C), updates to crew training procedures, technical problems with the hatch handle, inaccurate readings from the launch pad-fire detection systems and on the 27th the presence of unallowably high crosswinds. As Vaughan (1996) makes clear, such delays were normal rather than exceptional for shuttle launches.

7 NASA sub-contractor Rockwell also expressed doubts about the safety of launching on the 28th, due to the risk of ice hitting and damaging the shuttle or being aspirated by the Solid Rocket Boosters (Vaughan, 1996: 7).

of failure was downgraded into a very remote possibility. Key to understanding this process is the circulation of negative affect—anxiety—and its capture by technocratic confidence that took place across the teleconferences between NASA and Morton Thiokol.

Various studies (for example McConnell, 1987; McKenzie, 2001; Vaughan, 1996) have explored in detail the two teleconferences between NASA and Morton Thiokol that were held on the eve of the launch of Challenger to confirm approval to launch. All agree that what pervaded the second teleconference was a feeling of confidence that appeared to block some of the more anxious thoughts that had flowed between NASA and Morton Thiokol. The caution centred upon the effect of the sub-zero temperatures (predicated for January 28th) on the O-ring seals that are used to seal the segments that comprise the two solid rocket boosters. Several Morton Thiokol engineers predicted that the cold weather would harden the material properties of the seals sufficiently to create an ineffective seal between the segments during launch and allow the possibility of the 'blow by' of explosive gasses through the joint gap and the O-ring; this leak of hot explosive gas during launch could easily jeopardize the vehicle.

In fact, the resilience of the seals to cold had repeatedly concerned the engineers working at Morton Thiokol. These engineers had built up an archive of post-flight data which revealed evidence of blow by on cold launches, and induced anxiety within their team about the resilience of the joint. Thiokol had even been ordered by NASA, after 'blow by' evidence on STS-41B in April 1984 (launch temperature 57°F), to urgently review the design of the seals through laboratory testing. However, this review centred upon evidence that the application of putty within the O-ring, and its displacement through subsequent leak testing was the cause of the erosion—a theory which some evidence seemed to support. The findings of this review were delivered to NASA HQ in August 1985, and did not mention the effect of cold on joint, but did stipulate the need for escalating investigation of the problem. As a result, Thiokol initiated an internal O-ring task force. And yet, despite the mounting concern over the joint, both NASA and Morton Thiokol continued to confidently approve launches while a new design was developed (McConnell, 1987: 7; Vaughan, 1996: 299-305). These approved, and seemingly successful, launches of the Shuttle effectively blocked the transmission of anxiety within and around NASA about the seals. In fact, three of the five flights launched between the August review at NASA HQ and Challenger's final launch, experienced some degree of 'blow by.' But NASA appeared confident: past actions served as a guide for future events. The seals were being eroded but tolerance levels were said by NASA's SRB manager to be within acceptable thresholds; the evidence was not outside the 'data base' (until the eve of launch), and thus was seldom mentioned to senior managers (Rogers Commission, 1986: 148).

But this confidence was not purely subjectively relayed. Underpinning all of these discussions was another kind of confidence that the very existence of the seals produced. The O-ring seals were only required because the SRBs had to be broken into segments to facilitate transport via train from Morton Thiokol's plant in Utah, near the Rocky Mountains, to Kennedy Space Center for assembly, testing and launch. Before Morton Thiokol, rather surprisingly, won the SRB contract in the early 1970s, it was a small company. The contact award, and thus the seals, had raised questions of political nepotism between the then NASA Administrator James Fletcher (whose home state was Utah), Congress and the Utah political hierarchy (Klerkx, 2004: 86; McConnell, 1987: 7). Notwithstanding such criticisms, the design of the O-rings to contain hot gasses at launch was vital to ensure construction, test and reprocessing costs were minimized, and in turn that Thiokol as the cheapest bidder of four in the competition for the SRBs, and ranked lowest for technical quality, had won the contract (Rogers Commission, 1986: 120-1). Thus, the design of the seals, bound together far more than explosive rocket engines—they also provided a valuable conduit to circulate confidence in and around Thiokol, NASA, and the Shuttle.

Massumi (2002: 42) describes how 'confidence is the apotheosis of affective capture. Functionalized and nationalized, it feeds directly into prison construction and neo-colonial adventure' (p42) and now space travel. Confidence is defined here through a simple maxim: past action is a guide for future events; this closed orientation to the future is more than apparent within the systems thinking of NASA where spatio-temporal predictability and control are paramount (see Chapter 4). During the second teleconference, confidence reverberated loudly within NASA. For NASA managers, this feeling appeared partly as a bureaucratic survival tactic to avoid further funding cuts. NASA had to be seen to act confidently as this conferred the confidence of others. McKenzie (2001: 149-150) similarly describes an imperative of 'political accountability' dominating NASA prior to Challenger's launch. McKenzie explains this pressure as being part of a broader challenge for technological organizations to realize 'social efficacy' (p149-150). For Thiokol, such efficacy was essential to secure a renewed contract for SRBs.[8] Ultimately, for both institutions, confidence was politically valuable because it augmented the Reagan administration's preoccupation with the projection of national confidence, vis-à-vis the USSR, and by extension the American public (cf. Massumi, 2002: 41-2).

Within the teleconference, confidence blocked more open-ended, experimental possibilities for the future, not least the anxious negative affects released by Thiokol engineers Roger Boisjoly and Allan MacDonald, and their post-flight archive of O-ring 'blow by.' Both engineers remained deeply unconvinced about the performance of the seals in these frozen conditions. These engineers had gathered some evidence that the O-rings, a small but vital link in the circulation of the American transcendental state, were now no longer capable of circulating confidence or hope; instead they were inducing negative effects. If they were plugged into the launch, live on TV, these affects could be disastrously negative. However because they had been denied support to thoroughly investigate the

---

8   Indeed the Rogers commission that investigated the loss of Challenger revealed how Morton Thiokol's lucrative contract with NASA for the SRBs had been up for renewal when the decision was made to launch Challenger (Isikoff, 1986: A15).

resilience of the seals (see Rogers Commission, 1986: 141), Boisjoly could not put a definite number on his analysis of the risk. After all, no O-rings had actually completely failed before, but none of these launches had been subjected to temperatures below the 53°F encountered almost a year before on the flight of the space shuttle Discovery, STS-51C (Rogers Commission, 1986: 129-31). What was required was a test, but the means, and time, to perform this test were not available. As Vaughan (1996) puts it: 'the engineering argument failed because it did not meet the standards of scientific excellence prescribed by the original technical culture. In cultural terms, the engineering rationale for delay was a weak signal' (p398-9) And thus crucially, Boisjoly did not sound confident within the teleconference; rather he referred to the adverse launch conditions as being beyond 'goodness' (quoted in Vaughan, 1996: 317).

Managerial confidence, as both McKenzie (2001) and Vaughan (1996) describe, was well organized by NASA's bureaucracy: a set of prescriptive rules and procedures existed for identifying all manner of risks and making decisions based upon the transmission of measurable, scientifically verifiable, data up the agency's hierarchy (Chapter 4). Some of these checks were intended to gather data to prove that it is safe to launch confidently, as objective three of NASA's Shuttle Flight Readiness Review makes clear: 'Review solved problems and previous flight anomalies and establish *confidence* in solution rationale.' (quoted in Rogers Commission, 1986: 145; emphasis added). Remarkably, in the meetings prior to Challenger's launch, it now appeared proof was being asked that it was not safe to launch (Rogers Commission, 1986: 93). Bob Lund, Thiokol's Vice President of Engineering, later recalled:

> We had to prove to them that we weren't ready, and so we got ourselves in the thought process that we were trying to find some way to prove to them it wouldn't work, and we were unable to do that. We couldn't prove absolutely that that motor wouldn't work (testimony to Rogers Commission, 1986: 94).

This insistence on proving a negative, lead to a worrying situation: NASA's SRB middle managers had already denied their engineers the means to investigate, and potentially quantify a risk based on their confidence in successful launches; a risk that might emerge from unusually cold temperatures, and thus could *never* be *confidently* predicted in advance. This organization of confidence, thus confident organization, this insistence on predicting future events through a narrow range of past actions (successful launches in cold temperatures), meant that the affect of the cold on the O-rings, and thus on NASA, was blocked, at least within this meeting. Unlike Boisjoly, NASA's SRB project manager, Lawrence Mulloy, was exemplary in relaying confidence. Towards the end of the second teleconference, just before a 30-minute offline caucus for either side to decide whether to launch, Mulloy exclaimed, in defiance of the possibility of changing the predetermined launch parameters (the Launch Commit Criteria[9]) to include a stipulation not to launch below 53 °F: 'My God, Thiokol, when do you want me to launch, next April?' (quoted in Vaughan, 1996: 305). The answer from Thiokol when it came after the caucus was a confident one: 'MTI [Morton Thiokol Incorporated] recommends STS-51L launch proceeds on 28th Jaunary 1986. SRM [Solid Rocket Motor]-25 will not be significantly different than SRM-15 [the SRBs used on STS-51C].' Thus the experiment with the effect of the cold on the O-ring was the launch itself.

Almost 56 seconds after Challenger's launch, an unusually strong wind removed a build-up of oxides that had provided a temporary seal over a small gap in the O-ring seal in aft section of the right hand SRB; two seconds later the gap widened as a plume of hot gas emerged from the hole. Six seconds later the plume caused a leak in a tank of liquid hydrogen in the aft-section of the External Tank. Four second later and the Shuttle crew were given the all-clear by flight controllers to increase the thrust of the main engines to the pre-planned, 104 per cent of thrust[10]. Five second later: the hydrogen tank exploded and was forced into the forward oxygen tank; the right SRB broke from its mooring and collided with the External Tank; the External Tank exploded; the orbiter itself did not explode but as its launch trajectory was compromised, excessive aerodynamic pressure alone caused it to disintegrate. The stronger SRBs survived these forces, but were remotely detonated by launch controllers to avoid causing damage on the ground. Amid the disintegrating orbiter, the more robust crew compartment survived but now fell towards the Atlantic ocean from 65,000 feet (for a timeline of events see Rogers Commission, 1986: 20-39). At least three of the crew activated their emergency air supply and survived the 165minute fall, whether the crew cabin was sufficiently pressurized for the crew to remain conscious for the entirety of the fall and their eventual terminal collision with the ocean at over 200mph is not known (Kerwin, 1986).

The event was excessive to the rhythm of confidence in and around NASA, manifesting itself in a discernible shift in the NASA launch commentator, Steve Nesbitt's, tone from self-assured buoyancy to stuttered alarm: as he witnessed Challenger disintegrate, Nesbitt hesitantly responded: 'Flight controller here looking very carefully at situation. Obviously a major malfunction' (New York

---

9   The Launch Commit Criteria (LCC) is a set of standardized conditions devised by NASA in consultation with its contractors, which determine whether shuttle flight readiness can be approved. At the time of Challenger it relied heavily upon statistical data that demonstrated a reading within tolerance parameters that had already been identified in testing.

10   Although at launch the Shuttle's main engines operated at 100 per cent of thrust, at around 26 seconds after launch this thrust was gradually reduced to prevent the Shuttle over-speeding in the denser lower atmosphere, to limit damaging aerodynamic pressures. From around 60 seconds after launch the Shuttle passes into lower density atmosphere and the main engines were increased again. The thrust generated by the SRBs is constant during their planned 126 seconds of operation and cannot be halted once ignited at launch (NASA, 2013).

Times, 1986a; Sawyer, 1986: A1). Far from callous, Nesbitt's dry technical response was entirely in keeping with NASA's technocratic idiom. He simply could not render this affective event intelligible; it was excessive. The pause, the hesitancy, the affective suspense, of Nesbitt was repeated elsewhere. Unlike the Apollo 1 test, this event was witnessed live on television across the world, including by millions of American children, not least Christa McAuliffe's own class. NASA could not hope to control the release of negative affect by Challenger.

Massumi (2002) suggests that affective suspense—the inability to affect or be affected—always manifests itself in visceral, bodily, registers: 'viscerality is the perception of *suspense*. The space into which it jolts the flesh is one of an inability to act or reflect, a spasmodic passivity, so taut a receptivity that the body is paralyzed until it is jolted back into action-reaction by recognition' (p61)

And it is exactly this sense of affective suspense that Challenger seems to have released, at least initially. Joel Powell, a space enthusiast, later recollected what he saw on the day: 'I literally had no idea what was happening … The awful empty feeling and the anguished faces of the reporters and the lingering smoke cloud are something I will never forget. We were utterly helpless' (Powell, 1986: 184). The principal of the high school McAuliffe had worked at recalled her experience of watching the launch with the school's children—'We were rejoicing in the lift-off. We were exalting in it. We were celebrating with her. Then it stopped. That's all. It stopped' (Kiernan, 1986: A1). The experience of bodies being frozen in space and time was shared across American classrooms as one high-school student recalled on the front page of the *Washington Post*: after the accident 'I just stood there kind of empty for a minute.' Equally, the *Post's* reporter characterized the atmosphere in McAuliffe's classroom as 'horrified silence' (Vobejda, 1986: A1). The media reverberated with this tone of suspense, witness a *Washington Post* report from 29th January: 'The horror dawned slowly: for one very long moment after the explosion, few realized they witnessed a disaster' (Rensverg, 1986: A1). Evidence of this feeling of spatio-temporal suspense has also been found in academic analysis of the traumatic effect of Challenger, including children (for example Terr, 1992).

NASA appeared to try to contain this negative affect as it limited access to any information regarding the accident; not least the fate of the crew, pending an official investigation, just as it had done with Apollo 1. As the *New York Times* reported: 'Soon after the shuttle explosion Jan 28, NASA impounded virtually all information it had related to the shuttle program and instructed its employees not to provide information to reporters, even on an off-the-record basis' (New York Times, 1986c). NASA again sought to control the flow of information, and maintain the affective void of suspense. Just two weeks later, the *Washington Post* described how NASA's secretive approach contributed to 'NASA's decaying public image' (Johnson, 1986: A2). Mindful perhaps of the accusations of secrecy surrounding the Apollo 1 investigation, President Reagan inaugurated an independent investigation headed by President Nixon's Secretary of State, William Rogers. The Rogers Commission would take 120 days and involve 160 witness testimonies in live televised hearings; the Commission was instructed to investigate both the proximate and underlying causes of the failure and provide recommendations for improvement (see Rogers Commission, 1986, for a review see Vaughan, 1996).

Through the gathering of witness testimony, documentary and material evidence (including almost 90 per cent of Challenger), as well as test data, the Commission concluded that the O-rings, rendered ineffective by the ambient cold temperatures, had been the proximate cause of the disaster. This conclusion culminated in a memorably theatrical moment in the televised hearing on the 11th February when a piece of Shuttle O-ring material was being passed from Commissioner to Commissioner for inspection. Commissioner Richard Feynman, 1965 Noble Prize winning physicist, quietly asked for a cup of ice water for the lunch break. After the break he addressed NASA's Lawrence Mulloy, live on television:

> I took this stuff that I got out of your seal and I put it in ice water, and I discovered that when you put some pressure on it for a while and then undo it, it doesn't stretch back. It stays the same dimension. In other words, for a few seconds at least and more seconds than that, there is no resilience in this particular material when it is at a temperature of 32 degrees (quoted in Cook, 2006: 264).

This event penetrated the affective void induced by Challenger and perpetuated by NASA. The inflexible material was affected negatively by the cold, unable to circulate confidence, unable to respond at launch without diminishing other bodies, just as NASA. This experiment about the inflexibility of the O-ring material released negative affects into the void NASA had sought to maintain. This negative affect, analogous to Boisjoly's anxiety before launch—primed the media to report stories regarding different aspects of diminishment of NASA, and America. The day after Feynman's ad-hoc experiment, the *Washington Post* depicted NASA as a: 'a once-proud agency rudderless, uncertain, torn by dissension and low morale" (Johnson, 1986: A2).

The conclusion was inescapable: if NASA had performed this test themselves and those results had informed their decision-making before launch, Challenger would not have been lost. Just as with Apollo 1, NASA was shown to be organized to block utterly simple scientific evidence. Messianic hope and technocratic confidence in and around NASA had been diminished again by negative affect; this time released by an O-ring and a cup of ice water, live on television, replayed again in the evening news. Challenger's initial explosion induced affective suspense that primed America for an investigation that then released, via Feynman's experiment, negative affect: how could confident NASA, the beacon at the center of the 'shining city on the hill', be so incompetent, so incapable, so diminished? And why was this negative affect, now released through this high-school style experiment onto the global news media, not allowed to circulate in NASA prior to Challenger—how had it been blocked? The question the Commission now faced was how NASA had been organized in this way to be supremely confident, to forestall experimentation, to capture affect. Thus just like the Apollo 1 investigation, the Rogers Commission also examined NASA's organizational culture but this time on a far wider scale.

The Commission concluded that NASA was an internally divided organization, where middle managers, like Lawrence Mulloy, within Marshall Field Center were withholding potentially serious information, like the concerns over the O-rings resilience, from their seniors, and thus poor decisions were taken to commit to launch. The Commission concluded: 'This tendency is altogether at odds with the need for Marshall to function as part of a system working toward successful flight missions, interfacing and communicating with the other parts of the system that work to the same end' (Rogers Commission, 1986). McCurdy (1993) later identified this organizational conflict as bound up with a lack of political clarity (and funding) over NASA's mission, coupled to a bureaucratic-political diminishment of NASA's Apollo-era technical culture (see also Perrow, 1999: 379; Johnson, 2002; Vaughan, 1996: 396). Moreover, the Commission suggested this problem was long-established: the full extent of concerns of Thiokol's engineers about the O-ring within the Shuttle flight program were regularly withheld from senior NASA managers by NASA's SRB middle managers, not least their performance in cold conditions. The basis for the lack of communication with senior managers was partly said to be belief in redundancy (the O-rings had two seals) despite knowledge by senior managers within NASA that after launch the second seal could not function (Rogers Commission, 1986). Indeed this latter point—the confidence (and accountability) of senior NASA managers for the disaster is hardly discussed by the Commission—further evidencing what Vaughan observes as a culture of 'conformity not deviance' (p399) in NASA. Across NASA's hierarchy, evidence of past action was used as the basis for flagging up exceptional events. Once senior NASA managers had instigated the review of the SRBs in April 1984 (and subsequent presentation at NASA HQ in August 1985), the SRB middle managers in NASA believed there was no requirement to continue to inform their seniors of problems about the same issue. NASA even instructed Thiokol to request the problem be 'closed out', after becoming frustrated at its convoluted appearance during flight reviews with senior managers and its threat to launch confidence (Rogers Commission, 1986: 137-8). In effect, as the Commission report put it: 'NASA and Thiokol accepted escalating risk apparently because they 'got away with it last time' (Rogers Commission, 1986). Or, as Richard Feynman characteristically put it, they were playing 'a kind of Russian roulette' (Rogers Commission, 1986: 148; see also Burrows, 1998: 560). Contra Vaughan (1996: 380), this is more than mere conformity, it is confidence.

The Commission offered various recommendations to improve NASA including; technical improvements to the joint supported by more rigorous testing, overseen by the National Research Council (NRC); increased authority of program directors over Field Center operations; an independent safety and quality assurance Office within NASA; more thorough reviews of critical flight elements, audited by the NRC; and a cut in the Shuttle flight-rate to ensure time to resolve problems. What was being proposed was hardly a deviation from NASA's systems management.

In a concluding thought after the recommendations, the Commissioners then re-iterated the imperative that to foster confidence (through systems management) requires the release of messianic, hopeful affect. Specifically, NASA's capacity to engender a better America future elsewhere in space and time:

> The Commission urges that NASA continue to receive the support of the Administration and the nation. The agency constitutes a national resource that plays a critical role in space exploration and development. It also provides a symbol of national pride and technological leadership ... The Commission applauds NASA's spectacular achievements of the past and anticipates impressive achievements to come. The findings and recommendations presented in this report are intended to contribute to the future NASA successes that the nation both expects and requires as the 21st century approaches (Rogers Commision: 201).

The messianic hope, expressed here by the Rogers Commission, blended well into a wider climate of transcendental eulogizing for the crew of Challenger. Witness, President Reagan's eulogy two days after the disaster: 'They had a hunger to explore the universe and discover its truths. They wished to serve and they did—they served all of us' (Reagan, 1986). Tropes of an American transcendental state animated Reagan's eulogy as he compared the astronauts to 'pioneers,' wherein their deaths and our pain is 'all part of the process of exploration and discovery ... The future doesn't belong to the fainthearted. It belongs to the brave. The Challenger crew was pulling us into the future and we'll continue to follow them', as they 'slipped the surly bonds of earth to touch the face of God' (Reagan, 1986: A18). Reagan's religious eulogizing echoed remarks made a day earlier by Pope John Paul II: 'I lift up to God a fervent prayer so that he accepts in his embrace the souls of these courageous pioneers in progress of science and man' (Suro, 1986: page unknown). Here hope is drawn from within trauma, just as with Apollo 1.

For Reagan, these transcendental reveries primed his audience for Cold War ideological politics: 'We don't hide our space program, we don't keep secrets and cover things up ... That's the way freedom is and we wouldn't change it for a minute' (Reagan, 1986). Here, the affectively charged spectacle of the televised trials, NASA's own shocking diminishment, was rendered intelligible by Reagan as supreme verification of the messianic cause of America, despite NASA's own self-demonstrated departure from these ideals. The re-appointment[11] of James Fletcher as NASA Administrator continued this process of deriving hope from trauma—in a speech on his return, Fletcher argued: '[criticism of NASA] could do irreparable damage, not only to the agency, its people and its program, but also to the nation as

---

11   Fletcher had previously served as NASA's 4th administrator from 1971 to 1977, under Presidents Nixon, Ford and Carter. When re-appointed by Reagan in 1986, he was charged with overseeing a general improvement in reliability and safety after the loss of Challenger (Burrows, 1998).

a whole', because as Fletcher suggested, 'NASA is more than another government agency' it is 'a symbol of American aspiration and achievement, it is a vital national asset. Indeed, what NASA does deeply *affects* the way Americans look at ourselves and the way the world looks at us' (Fletcher, 1986: 1; emphasis added). Yet, despite the confident posturing, Fletcher knew NASA was an institution enacted through political (and technical) compromise as much as confident dreaming as his days working with Nixon made clear (Klerkx, 2004).

In marked opposition to the confident oratory of Reagan and Fletcher, the Soviet news media, cited the urgency of NASA, partly fuelled by Kennedy's messianic hope in space exploration, to critique the image of America as the transcendental state. For example, a Polish newspaper, *Zolnierz Wolnosci*, explained the disaster as evidence of imperialist ambition to militarize space wherein 'seeking quick progress in militarizing space, created a climate of haste and nervousness at NASA, compromising safety' (quoted in New York Times, 1986b). In a similar vein, the Communist Party newspaper *Pravda* noted how 'Although it seemed that the tragic spectacle of the public death of people should underline the extreme need to unite man's creative efforts in the difficult cause of the peaceful use of space, in Washington there are officials who find it possible to call for the speedy militarization of outer space' (quoted in Dee, 1986: A12). On 30th January 1986, Russian news agency *Tass* explained how, and not entirely without foundation (see Chapter 6), that Challenger exploded because NASA 'was in a hurry to turn them [shuttle flights] into real commercial flights' (quoted in New York Times, 1986b). A day later, a Russian television news reporter remarked that space technology 'remains a world of terrible unpredictability, even though politicians speak about some kind of brilliant triumph ... of security and confidence' (Fesunenko, 1986). On the same day the Russian new agency *Tass*, starkly suggested that space technology is 'a weapon that is capable of blowing up the world and which is beyond their [politicians] control however hard they try' (Tass, 1986). In August 1986, Moscow television news went further and criticised how Reagan had cleared the Shuttle's schedule to focus on military missions instead of commercial payloads[12]: 'it has once again shown that in conditions of capitalism, technological progress nearly always turns against man and his vital interests ... Technological progress has been put to the service of the most reactionary forces of militarism' (MDS, 1986).

As journalists and politicians attempted to render the affect of the event intelligible within sociological, technical or ideological registers, something less

---

12   President Reagan announced this shift in space policy towards the end of June 1986, in part because the aerospace industry was unable to compete with the government subsidies that kept the price of shuttle cargo down. The decision finally put pay to the notion, put forward in the 1970s, that the shuttle would make space travel profitable, an idea that had helped promote the shuttle program; it also made available more shuttle launches for military and scientific missions (Tapscott, 1986). The payload on STS-51L was a civilian communication satellite and scientific instrumentation.

Figure 8.2   Orbiter and external fuel explosion and separated SRB exhaust trails, 28th January 1986, Kennedy Space Center Florida, NASA image: GPN-2004-00012

Source: NASA

intelligible but perhaps more affective remained—the shock and suspense of the event itself was replayed in images of the explosion found in television and film. Images of Challenger exploding (Figure 8.2) performed a similar role as they were re-used in the print media. In 1996, ten years after the disaster, the *Washington Post* remarked the image was 'used to illustrate, if not actually embody, a whole succession of ideas' (Washington Post, 1996: C5), from NASA's 'bureaucratic arrogance' during the accident investigation, to the 'political evil' of the White House and Congress, which had pressured NASA to compromise design quality in the early 1970s (Chapter 6), but above all it, concluded the image 'presented the Challenger explosion as a dramatic moment in an extended historical process: American decline' (pC5). As evidence for the final point, the *Post* describes how it was used in visual montages of American decline, 'because it was so dramatic it was often the rhetorical exclamation point to such visual arguments' (pC5). What is absent from the analysis offered by the *Post* is acknowledge that the image arrived first, the shock, the sounds all arrived before we could contemplate its significance, its meaning. This image does not simply engage us but rather it primes us to be shocked, to become anxious, and then perhaps hopeful of a better future, even confident, just like Reagan.

Yet the meaning of this image is far from easily contained with a nationalistic script. Witness two examples of its re-use after the disaster in adverts placed in the national news media. The first was an advert placed in the *Washington Post* in 1987 by the Society of Professional Journalists and the Advertising Council; it contained three images (one of the Shuttle disintegrating, one of launch with the plume from the right SRB circled and one of the frozen launch pad), using the slogan 'If the press didn't tell us, who would'. The second was an advert placed in 1990 by the Government Accountability Project, a non-profit making organization, encouraging and protecting government and corporate whistleblowing. In the advert, beneath an image of the disintegrating Shuttle, the organization promotes itself as making 'government honest, open and accountable' (GAP, 1990).

Such explicit attempts to circumvent the capture of shock and anxiety by the State through confidence and hope did not appear to engage the public in large numbers and invoked angry feedback from some in NASA (Washington Post, 1987). Instead, Challenger ultimately appeared to actually help circulate Reagan's renewed confidence in NASA, and hope in the exceptional transcendence of America in space and time. Indeed, a survey of public opinion by the Northern Illinois University (NIU) Public Opinion Laboratory in 1987 (NIUPOL, 1987), concluded that 95 per cent of its respondents believed the Shuttle was still an outstanding piece of American technology that should be fully supported; and that it failed because of its complexity rather than 'bureaucratic arrogance' or 'political evil'. The NIU report concluded: 'the short-term effect of the Challenger accident was to increase the already strong sense of national pride in the shuttle program and to stimulate a significant sense of personal loss in regard to the seven astronauts killed' (NIUPOL, 1987: 50). Reagan's release of messianic hopeful affect within narratives of heroic sacrifice is surely crucially in priming such thoughts.

Similar imagery of transcendental rebirth accompanied the restart of America's human spaceflight program on the 29th September 1988. Two years after the disaster, the Shuttle Discovery (STS-26) was launched from Kennedy Space Center; this time the crew were wearing full pressure suits, equipped with bail-out systems, and were powered by SRBs incorporating the more resilient joint design. On the fourth day of the flight, live on television from orbit, the astronauts paid testimony to their own and the nation's emotional journey. Mission Specialist, David Hilmers reflected: 'Many emotions well up in our hearts—joy, for America's return to space—gratitude, for our nation's support through difficult times—thanksgiving, for the safety of our crew—reverence, for those whose sacrifice made our journey possible' (quoted in NASA, 1988). With even more patriotic lyricism, Mission Specialist, John Lounge emphasized the national need to explore: 'we are convinced that this is the road to the future—the road America must travel if we are to maintain the dream of the Constitution—to secure the blessings of liberty to ourselves and to our posterity' (quoted in NASA, 1988).

As the Rogers Commission indicates (see also Burrrows, 1998: 560), confidence, that is, the capture of negative affect, no doubt contributed to the losses of Apollo 1 and Challenger (and other spacecraft and crews). Vaughan (1996), in her sociological investigation of the normalization of deviance in NASA, is suggestive of this point as she concludes: 'What is compelling is how structures of power, history, processes, and layered cultures that *affected* all participants' behavior at a subtle, prerational level combined to produce the outcome' (p399; emphasis added).

My analysis here elaborates upon these proposals that some kind of pre-cognitive, but systemic, force was at work in NASA. I suggest that this force was affect. Specifically, the organized capture of negative affect by confidence in and around NASA, and the accompanying release of positive affect: hope. However, forensic diagnosis of the (affective) relays involved in these disastrous events is not my concern *per se*. Rather my purpose in this chapter is to understand how the hopeful edifice of the American transcendental state was repaired (re-territorialized) after it was disorganized (de-territorialized) as it passed through two unruly conduits: faulty electrical wiring in the Apollo 1 vehicle and a leaky O-ring joint in the right SRB of Challenger.

But these two events did not operate in the same way. Unlike the loss of Apollo 1, Challenger was a shared media spectacle—a place (for most) to (safely) experience negative affect, perhaps in order to become primed for heroic, ideological, or religious, recognition, for confidence and hope (Penley, 1997: 48-49). The philosopher of science, Michel Serres, has perhaps gone the furthest in making this argument as he compares the loss of Challenger, and our repeated enactment of it via television, to the worshipping of a Carthaginian funeral pyre[13]; thus modernity is connected to its ancient past of sacrificing human life to the heavens:

> religion is in technology; the pagan god is in the rocket; the rocket is in the statute; the rocket on its launching pad is in the ancient idol—and our sophisticated knowledge is in our archaic fascinations. In short, the construction of a failed or successful society is in the successful or failed project of going toward the stars (Serres, in Latour and Serres, 1995: 160-1).

---

13   Specifically, the 'ancient Carthaginian practice of enclosing humans [including children] in a gigantic brass structure of the god Baal and incinerating them there, as a sacrifice to their deity' (Serres and Latour, 1995: 205). For Serres, both this practice and Challenger can be regarded as a cosmic sacrifices because they involve: the worship of massive, technological objects; voyages beyond Earth; and crucially, denial—in Carthage parents denied the screams of their children were human, for Challenger this denial is an insistence the loss was an accident, even though, as Serres (in Serres and Latour, 1995) explain, it was 'inevitable, even calculable, through probabilities' (p160; see also Penley, 1997: 48-9).

Despite the controversy surrounding this comparison (see Serres and Latour, 1995: 205), it is difficult to disagree that Challenger induced widespread, and repeated, affective engagement: a 2001 US survey of the most closely followed news events of the previous twenty years placed the Challenger story top (Anon, 2001). Unlike Apollo 1, Challenger could be repeated via the printed and visual media; and thus negative affects like shock, trauma and anxiety, were harnessed to dis/organize (that is 'de/re-territorialize') bodies far beyond NASA, including; Cold War politics, NASA's funding, American technocracy, systems management, media openness, and school children's desires to become astronauts. Yet, there are also clear similarities between the affective patterning of these two events: most significantly, these two catastrophes throw into relief mechanisms through which negative affects (shock, anxiety and trauma) appear both the *cause* and *effect* of technocratic confidence in and around NASA, and messianic hope in the American transcendental state. And this is precisely how, and why, on the 28th September 1988, on the eve of the launch of STS-26, Shuttle Commander Frederick Hauck reflected: 'Dear friends, we have resumed the journey that we promised to continue for you; dear friends, your loss has meant that we could *confidently* begin anew' (NASA, 1988).

Chapter 9
Critical Cosmopolitics

In the preceding eight chapters I have argued that some of the unique qualities of outer space—vastness, Otherness, sublimity, timelessness, spacelessness—are just as integral to extra-terrestrial projections of US geopower, as its well-known capacity (Arendt, 1963; Cosgrove, 2001; Dickens and Ormrod, 2007; Dolman, 2001; Macdonald, 2007) to function as an Archimedean high point to monitor and control the surface, and atmosphere, of the Earth. While the focus of my study has been the United States, and more specifically NASA, the implications of this cosmic projection of geopower—the American transcendental state—are global in reach, from enabling and shaping imperialistic ideologies (Chapters 1-3 and 7) to fuelling the extension of technocratic managerialism (Chapter 4-6 and 8). What is more, messianic hope in America remains a global commodity, consumed, for example, through the internationally franchised *Star Trek* television episodes and films (Penley, 1997: 98-99), multinational 'Space 2.0' corporations, like SpaceX (Chapter 6), worldwide audiences to the addresses of American presidents (Chapter 6) and global tourist attractions like the National Air and Space Museum and Kennedy Space Center Visitor Complex (Chapter 7). These global circulations suggest that while my empirical focus in this study has been on the extra-terrestrial assemblage of the American transcendental state, as viewed from within the borders of the US, the salience of my analysis is *geo*-political.

The development of the American transcendental state through space exploration must also be viewed as an integral component of a far older geopolitical project—the production of an American identity defined in terms of the transcendence of limits, whether technological, economic, spiritual or territorial, enabling the moral aggrandizement of the past, present and future of a horizontal strata of sovereign territory and its peoples (McDougall, 1997; Noble, 2002; Nye, 1994; O'Brien, 1988; Ricard, 1999; Stephanson, 1995). Over the last decade or so, a growing number of scholars, including geographers, have turned their attention to how messianic-exceptionalist visions of America as the 'Promised Land' of 'Chosen People' have inflected various imperialistic projects including: the pursuit of democracy through military intervention in the 'global south' (Anthony, 2008); the technocratic 'greening' of Western global capitalism (Singer, 2010); the building of a 'culture of war' in foreign policy (Marsella, 2011), the circumvention of international institutions (Agnew, 2006); and most prominently perhaps, George W. Bush's 'war on terror' where invasions of Afghanistan and Iraq became justified as a 'cosmic struggle between good and evil' (Agnew, 2006: 183; see also Barkun, 2010; Dijink, 2006; Strum, 2010; Wallace, 2006). All of this work indicates two points: first, the enduring Apocalyptic influence of dispensational pre-millennialism[1] on both interventionist and isolationist currents within American (geo)politics (Strum and Dittmer, 2010: 18); and secondly, the rise of a religious cosmology that positions America at the moral, geographical, and spiritual, centre of the universe (Strum, 2010: 150).

My analysis of American spaceflight adds to this body of work on religion and geopolitics by drawing attention to five less discussed conduits of this pious vision of American geopower: (i) *the secular*—museums, family theme parks, systems management; (ii) *the sublime*—astronomical artwork, Moon landings and distant Nebula; (iii); *the profane*—Nazi slave labor camps, technocratic patriarchy, and dead astronauts; *the technological* (iv)—rocket production lines, O-rings, electrical wiring; and (v) *the revolutionary*—female astronauts, May 1968, and Richard Feynman. Analytically, these diverse registers suggest the utility of working with a broader, less explicitly spiritual, set of theoretical assumptions, to address the cosmological aspects of American geopolitics. This is why I mobilized the concept of the 'American transcendental state', rather than 'deified nation' (O'Brien, 1988: 41) within this study. This deliberately hallucinogenic sounding term captures some sense that the messianic-exceptionalistic projection of American geopower is a more diffusive, experimental, fantasmic, embodied, and ostensibly secular, affair, than conveyed within much discursive analysis of the religious undercurrents inflecting American geopolitics (for example Agnew, 2006; Dijink, 2006; Strum, 2010; Wallace, 2006).

I would like to suggest now that there is another benefit in bringing together these diverse practices under a broader analysis of the American transcendental state: their common geography becomes all the more obvious. That is, all these practices involve thinking, doing or resisting, celestial transcendence as an apparatus of American geopower; hence they can all be rightly considered 'vertical geopolitics' (Elden, 2013; Graham, 2004; Graham and Hewitt, 2013). This label has developed to identify a body of work addressing how the circulation of American geopower involves more than two-dimensional geographies of area. It currently includes analyses of; drone warfare (Gregory, 2011); aerial bombardment (Graham, 2004); police helicopters (Adey, 2010); satellite surveillance (Macdonald, 2007) and satellite drone navigation and targeting (Gregory, 2011). Elden (2013: 40) explains that 'vertical geopolitics' is mostly focussed upon how state political

---

1   Strum and Dittmer (2010) identify 'dispensational pre-millennialism' as the most influential thread of Judeo-Christian eschatology within American (geo)politics. This popular subset of high Protestant belief, while far from unified, tends to share certain common ideas: (i) Christians and Jews are 'Chosen People' from 'Promised Lands'; (ii) life on Earth can be divided into epochs or 'dispensations' (iii) Jesus and the anti-Christ will arrive on Earth in the next epoch—'Christ's millennium'; (iv) Upon his arrival, God will judge humanity during an apocalyptic war between Jesus and the anti-Christ called 'tribulation'; (v) Jews and Christians will be spared from suffering through the 'rapture'— the raising of souls from danger into Heaven; (vi) Christians will inherit Heaven and the Jews will inherit the Earth—building a 'new Israel'—purged of evil.

technologies allow diverse populations to be measured, calculated, controlled and killed, 'from above', and occasionally 'from below' (for example Elden, 2013; Graham and Hewitt, 2013). By contrast, the vertical orientation I have adopted here, while related, is different. Specifically, I have described how aspects of the projection of American identity, geopower, and territory, *also* involve a vertical spacelessness—a deterritorialization—a potential collapse into sublime, cosmic, insignificance; in short, rather than the 'view from above', the perspective I have traced has been a 'view into the above' (and back). In part, therefore, my study can be considered a response to Elden's (2013) recent question: 'How would our thinking of geo-power, geo-politics and geo-metrics work if we took the earth; the air and the subsoil; questions of land, terrain, territory; earth processes and understandings of the world as the central terms at stake, rather than a looser sense of the 'global?' (p49)

I propose we add to this list celestial entities, including the Moon (Chapter 3), the Martian surface (Chapter 6) and the Eagle Nebula (Chapter 7), as well as God (Agnew, 2006; Dittmer and Strum, 2010; Strum, 2013). Thus, perhaps we should be cautious of Elden's (2013b) rather geocentric call 'about how geopolitics might be thought as earth-politics rather than simply a synonym for global politics' (p59). Instead, it might be more useful to bear in mind Deleuze and Guattari's (1988: 101) argument that even absolute deterritorialization—something akin perhaps to the mathematical cosmic sublime of Kant (Nye, 1994: 7-8)—always involves reterritorialization(s). Recall how Charles Bonestell (Chapter 2), William Clancey (Chapter 6) and the National Air and Space Museum (Chapter 7), respectively, and persuasively, associated vistas of the Moon, Mars and the Eagle Nebula with the American West, and by extension locate America at the centre of God's universe (Boime, 1991; Stephanson, 1995).

This analysis of American spaceflight also sheds light on seldom acknowledged connections between religious and vertical geopolitics and technocracy. The relation between critical analysis of geopolitics (O Tuathail, 1996) and technocratic management (Alvesson, 1987), remains remarkably undeveloped. Arguably this lacuna says more about the disciplinary separation between critical security studies and organization studies (Grey, 2009) than the various intellectual cross-fertilizations between organization studies and human geography (Clegg and Kornberger, 2006; Dale and Burrell, 2008; Parker, 2013). Nevertheless, there are, as Grey (2009) maintains, clear resonances:

> Indeed it could said that, in the same way that the development of security studies in particular, and organization studies to an extent, was shaped by geo-politics of wars both hot and cold, so too many current and future directions be in part a reflection of developments in contemporary geo-politics (p31).

Some organizational practices are of course, very much on the 'front line' of practical geopolitics; that is, they comprise the 'the foreign policy bureaucracy' (Ó Tuathail and Dalby, 1998: 4) through which geographical concepts are deployed to aid 'conceptualization and decision making' in 'everyday foreign policy' (O Tuathail, 1999: 110). Examples here include the work of the US Air Force, the CIA (Central Intelligence Agency) and the UK's Foreign and Common Wealth Office. There are also a host of other organizations that no doubt influence how practical geopolitics is produced, from security analysts like the RAND Corporation to global defense contractors like McDonnell Douglas. However, analysis of the relationship between organizational and geopolitical practices remains embryonic. For example, Anderson's (2011) study of urban counter-insurgency and Gregory's (2011) of drone warfare, do no more than merely infer that the rise of the 'networked organization' is reworking the projection of American geo-power. Correspondingly, two organizational studies of the military only hint that, for example, masculine discipline (Godfrey et al., 2012) and team identities (Corona and Godart, 2010) shape and are themselves shaped by grand geopolitical narratives like the 'war on terror'.

But the imbrication of geopolitical and organizational practice can also be more subtle and much less militaristic—concerning the anticipation and cultivation of geopower through shared national identities, that is 'popular geopolitics' (O Tuathail, 1999: 110). Here, the connection to organizational practices is no less significant, yet invisible in the literature. NASA offers a good example: from its inception, the space agency developed increasingly refined technocratic techniques that aligned people and machines to naturalize the pursuit of a popular geopolitics wedded to American geopower. Viewed in this way, imperialistic geopower and technocratic-managerialism are interwoven forces; hence the present study suggests the richness of more sustained critical analysis of organization and geopolitics.

However, I am all too aware that in stressing the widespread application of this concept of the America transcendental state to understand American geopower— and, concomitantly, the fecundity of bringing together analyses of religion, verticality and now technocracy within critical geopolitics—I run the risk of constructing a totalizing, monstrous, edifice. The reader might rightly ask at this juncture, paraphrasing Nietzsche, have you not gazed into the cosmic abyss of American geopower for too long; are you not also reifying American geopower in the cosmos rather than challenging it? Indeed, throughout the preceding chapters I made reference to a rather singular sounding concept of *the* 'American transcendental state'. But, as in the introduction, I must stress again here, that I took this decision for reasons of analytical clarity rather than to suggest I have revealed an *independent, singular, definite* and *a priori* reality (Law, 2006), some essence akin perhaps to what Agnew (2006: 184) refers to as 'Americanism'. Instead, within each chapter I have traced the progressive assemblage of the American transcendental state—that is, nothing less than the divinely sanctioned, exceptional, and messianic, right and duty, of America, and its leaders in its name (Wallace, 2006: 225), to command cosmic space and time by evoking forces of 'good' and 'evil', 'us' and 'them' (Agnew, 2006; Strum, 2010). But the immutability of this cosmic vision (Strum and Dittmer, 2010; Wallace, 2006) belies the transformative, fragmented, heterogeneous components that sustain it, across landscape artwork, through Kennedy's Moon Speech, to the O-rings of Space Shuttle Challenger. Throughout this study I have suggested countless relations through which this vision is not only produced (Dijink, 2006; McDougall, 1997; Noble, 2002; Nye, 1994; Ricard, 1999; Stephanson, 1995; Wallace, 2006) but circulated, maintained, resisted, repaired, transformed, and experimented with.

How then to conceptualize this heterogeneous, but obdurate, cosmic being? Latour's actor-network theory (1987; 2005; 2012) is useful to an extent here; first, we can conceptualize the transcendental state as an 'immutable mobile' that 'ends up traversing the universe' by 'pay[ing] for each transport with a transformation' (Latour, 2013: 127); it is 'not displacement *without* transformation but displacement *through* transformation (Latour, 2005: 223); second, the transcendental state can be understood as offering a prophetic, but partial, 'panorama' of the 'world [cosmos] to be lived in' (p189) which must then, in turn, be:

> ... carefully situated inside one of the many Omnimax theatres offering complete panoramas of society—and we now know that the more thrilling the impression, the more enclosed the room has to be. [American] Society is not the whole 'in which' everything is embedded, but what travels 'through' everything, calibrating connections and offering every entity it reaches some possibility of commensurability. (p242)

Read against Latour's concepts of the 'immutable mobile' and the 'localizable panorama' it is easy to see why my analysis of American transcendental state has involved mapping circulations within as well as beyond our lives. And this is a political move too, because it suggests that opportunities to test and resist the American transcendental state are closer to hand than we might think. As revealed in Chapter 8, a great deal of effort is required to keep the transcendental state circulating because the heterogeneous conduits it passes through—electrical wiring, teleconferences, flight readiness reviews, budget decisions and O-ring joints—are capricious and experimental; that is, affective. Other Chapters acknowledged similar fragility accompanying the assemblage of the transcendental state, including; the partially-owned Declaration of Independence (Chapter 1), the globally unifying *Earthrise* photograph of Apollo 8 (Chapter 3) and the rusting rockets on display in the gardens of the Kennedy Space Center Visitor Complex (Chapter 7). Now located within this chain of heterogeneous transformations, what strategies might aid us in purposefully transforming this now confined totality? Or put differently, how might we engage outer space to resist this cosmic deification of America (O Brien, 1988)? In concluding this study, I propose three techniques but no doubt there are many more.

First, we can expose the void at the heart of this messianic-technocratic projection of geopower (Wallace, 2006). This approach was evidenced in Chapter 1 by Derrida's (2002) deconstructive reading of Declaration of Independence. Derrida (2002) emphasizes how signing the Declaration in God's name entails no democratic ownership over America's future, in outer space or elsewhere. Across the development of American spaceflight, the perils of messianic, free-floating, notions of 'Progress', 'Exploration,' 'Frontier' and 'The Future' are all too apparent, not least for NASA itself. Lester and Robinson (2009) suggest the emergence of this critique within the American space policy community:

> We should accept that "exploration" is a multivalent term, with many meanings, some of which are contradictory, and all of which have historical precedent. For too long we have looked at the history of exploration selectively, seeking to find the antecedents which justify our own vision of exploration: as science, as human adventure, as geopolitical statement. This is a definitional fight which cannot be won. Space policy must acknowledge the multiple visions for space exploration, developing a clear-eyed metric of value which avoids the vagaries of lofty "exploration-speak", If the merits of human exploration of the Moon and Mars are primarily symbolic and geopolitical, what are these goals worth in terms of federal funding?

I am unconvinced by the economically instrumentalist conclusions made by Lester and Robinson (2009) about putting a value upon even NASA's 'softer' geopower, but the general caution about harnessing nebulous messianic mythologies to advance American space exploration is valuable. Of course the problem is this tradition of finding our God in the cosmos is long-established as Olsson (2007) suggests via this retelling of the Babylonian creation epic, *Enuma elish*:

> Marduk is the Lord of lords ... Hail to the Chief! Fifty were his names, so numerous that if ever attacked he could always hide behind another alias. Never catchable as the specific this or that, always on the move as an ambiguous this and that ... Ungraspable multiplicity. ... In this mist-enveloped region of religion naming is the name of the game, an exercise in ontological transformations where earthly people appear as projections of heavenly gods, social relations as signs in the sky. ... a signified meaning searching for its own coordinates (Olsson, 2007: 23).

Perhaps a more modest approach is required: we should simply resist the urge to name, and tame, the cosmos as a Whole, by naming a celestial Godhead in it that we claim for ourselves (Wallace, 2006) but cannot ever fully own. 'Evil is the disaster of a truth when the desire to force the naming of the unnameable is unleashed . .... Evil is not disrespect for the name of the other, but rather the will to name *at any price*' (Badiou, 2004: 115-6; original emphasis). Challenging the cosmic aggrandization of America might therefore imply some attempt to resist naming *our* God/Future/Progress in the cosmos. Put simply, this all too easy act of cosmic de/reterroritalizaiton is too crude, too undemocratic, too costly.

A second, related, strategy which can be adopted to resist the American transcendental state was discussed within Chapter 3; this is the capacity to push transcendence to another plane or refuge—to follow one line of flight of cosmic deterritorialization and then re-territorialize the Earth in a panorama that starts with a common human experience, rather than those of any particular nation/ God/future. The aim of this strategy is to mobilize a cosmic imagination that can register something of the shared experience of being human.

In Chapter 3 I discussed how the *Earthrise* photograph from NASA's Apollo 8 mission have stimulated new cosmic imaginations—including 'spaceship' Earth (Cosgrove: 2001, 257-262; Henry and Taylor, 2009; Ward, 1964), Noetic science (Benjamin, 2003: 60-61), global political ecologies (Connolly, 2002)—that defied nationalistic appropriations by inferring a human transcendence. However, as the American author Kurt Vonnegut explains such a transcendental image of humanity, emptied of territorial divisions and difference, is not itself without risk: 'Earth is such a pretty blue and pink and white pearl in the pictures NASA sent me. It looks so clean. You can't see all the hungry, angry earthlings down there—and the smoke and sewage and trash and sophisticated weaponry' (Vonnegut cited in Burrows, 1998: 423). Similarly, Deleuze and Guattari (1988) suggest we should always remain sceptical that de-territorialization is a progressive act on its own: 'Never believe that a smooth space will suffice to save us' (p500).

A third strategy is to augment different affects amid the assemblage of the American transcendental state. As described in Chapter 8, the American transcendental state depends upon the cultivation of confidence in technocracy allied to an affective becoming hopeful—a positive openness to the future as life enhancing—orientated around the transcendence of America in cosmic space and time. But, as Anderson (2006), explains, becoming hopeful does not necessarily need to operate in this transcendental manner: hopefulness can also emerge not to ward off suffering, but through every day sorrows, through diminishment of the body's potential to affect and be affected. Consider, for example, how Dotty Duke refused to discuss her fears and anxieties with her astronaut husband as she kept the 'house in order and [took] out the garbage' (Duke 1990—Chapter 5). Dotty Duke epitomizes a different kind of becoming hopeful—a capacity to remain open-ended about the future in a life enhancing manner through diminishment—devoid of discussion of a better future in Earth or in the cosmos; this is hope that challenges 'the easy equation between transcendence and a future elsewhen or elsewhere in favor of an imminent transcendence from within vectors of diminishment' (Anderson, 2006: 749; for more analysis of immanent transcendence related to Space see Smith, 2009: 211).

Another affect which is useful in short-circuiting the hopeful assemblage of the transcendental state is boredom. Anderson (2004) describes boredom as the moment when the '"forgetting" intrinsic to habit has been momentarily incapacitated. It is the unravelling of habit, a sudden realization of the again' (p743). Boredom depresses the life enhancing capacity of ourselves to be open to the future, engendering stillness and slowness of thought-action in space-time, where, as Anderson (2004) puts it, the capacity to experience the 'not yet' (p749) is suspended. The evolution of American spaceflight might appear to some the antithesis of boredom, but, as Jorgensen (2009) suggests, the American humanization of outer space has gone hand in hand with endless repetition (of middle America):

> The August 1969 Life Special Issue, released to commemorate the landing, wants to produce sympathetic accounts of the astronauts. It is filled with glossy, high color photographs of the astronauts not only mastering outer space, but their domestic spaces as well. Neil Armstrong bakes pizza, Buzz Aldrin jogs through the suburbs, and Mike Collins prunes his garden. These images resonate with outer space itself, as the astronauts use tools in both terrestrial and extraterrestrial environments. The spatula and shears the astronauts use to cook lamb curry and prune roses with resemble the objects they hold while walking the moon, these being a laser reflector, seismometer and solar wind sheet (p179).

There is no hopefulness on offer in Jorgensen's (2009) reading of American spaceflight. Instead the boredom experienced in the cosmic repetition of middle America signals despair: 'Apollo 11 represented an America that had become unhinged by its own technocracy, its middle class lifestyle, and television' (p188). Jorgensen (2009) is not, of course, alone in identifying aspects of spaceflight repetitive, even boring. As the emergence of the Teacher in Space program demonstrated (see Chapter 8), NASA itself has historically attempted to introduce elements of excitement, even increased risk, to engage a global audience. Yet, of course, a balance has always had to be struck, as Parker (2009) explains of Apollo: 'Everything was supposed to be boring, because boredom meant no surprises, and hence the possibility of the adventure in some sense rested on its denial' (p326). Although fleeting, boredom is surely an unavoidable ingredient in NASA's technocratic confidence, but when focused and channeled, it does suspend hope in the cosmos as a better place, perhaps providing an opportunity for us to pause and register something of the sublime Otherness of Space, where we concurrently repeat and differ ourselves into infinity: 'Media representations of space travel turn the vastness of space into the similitude of domesticity, as human familiarity comes to stand in for the infinite. At the same time, the domestic attains the dimensions of the infinite, and in turn becomes strangely unfamiliar to the television viewer' (Jorgensen, 2009: 179).

These three techniques of *cosmo*-political intervention—refusal to name, human transcendence, and sensitivity to new affects—are all worthy of greater attention, especially when they can be connected up to, and interfere with, the assemblage of the American transcendental state. Clearly not all of those involved directly in the development of spaceflight will want or be able to practise these techniques. Nevertheless even among this group these techniques are intended to offer greater receptivity to new cosmographical imaginations which move beyond the cosmic aggrandization of messianic-imperialistic-technocratic impulses. If we have entered the Cosmic Age where all territorializing assemblages, all States, now derive vital energy from the Cosmos (Deleuze and Guattari (1988: 342), then the imperative becomes not to simply do cosmopolitics (Latour, 2005) but rather which *cosmo*-politics do we want to pursue? My favoured vision of a Geography of Space is one where this question is endlessly asked but never answered with absolute confidence.

References

Ad Astra (1991) 100 stars of space: the Challenger Seven, *Ad Astra*, July/August edition, p. 13.

Adey, P. (2010) Vertical Security in the Megacity Legibility, Mobility and Aerial Politics, *Theory, Culture & Society*, 27(6): 51-67.

Agnew, J. (2005) *Hegemony: The New Shape of Global Power*, Temple University Press: Philadelphia, PA.

Agnew, J. (2006) Religion and Geopolitics, *Geopolitics*, 11(2), 183-91.

Allen, C. (2004) *The Udvar-Hazy Center, Technology and Culture*, 45(April), 358-62.

Alsop, S. (1957) Sputnik May be ICBM's Eyes, Matter of Fact Article Series, *Washington Post*, 13th October 1957.

Anderson, B. (1991) *Imagined Communities: Reflections on the Origin and Spread of Nationalism*, Verso: London.

Anderson, B. (2004) Time-stilled space-slowed: how boredom matters, *Geoforum*, 35(6), 739-54.

Anderson, B. (2006) Becoming and being hopeful: towards a theory of affect, *Environment and Planning D: Society and Space*, 24(5), 733-52.

Anderson, B. (2011) Population and Affective Perception: Biopolitics and Anticipatory Action in US Counterinsurgency Doctrine, *Antipode*, 43(2), 205-36.

Anderson, B. (2012) Affect and biopower: towards a politics of life, *Transactions of the Institute of British Geographers*, 37(1), 28-43.

Anon (1968) *Space Daily*, June 3rd, p. 156.

Anon (1971a) Expert's vain warnings on deadly Apollo 1 told, *Los Angeles Times*, September 13th.

Anon (1971b) The capsule fire flares up again, *LIFE* 72 (12), September 17th.

Anon. (1957a) Grave Defeat to America, *Herald Tribune*, 5th October 1957.

Anon. (1957b) World Newspapers See Soviet Taking Lead From United States in Space Science, *New York Times*, 7th October 1957.

Anon. (1957c) Conquest of Space, *Washington Post*, 7th October 1957.

Anon. (1957d) Russell says Satellite Confirms Reds Ballistic Missile Claims, *Washington Post*, 6th October 1957.

Anon. (1967a) Apollo Study Stirs Critics, *Baltimore Sun*, April 10th, p. 5.

Anon. (1967b) Source says Spacemen Didn't Die Immediately, *The Evening Star*, Washington DC, January 11th.

Anon. (2001) What's News?, *National Journal*, 30th June.

Arendt, H. (1963/2007) The Conquest of Space and the Stature of Man, *The New Atlantis*, 18(fall), 43-55.

Avery, K. J. and Harvey, E. J. (2003) *Hudson River School Visions: The Landscapes of Sanford R. Gifford*, Metropolitan Museum of Art Press: New York.

Badiou, A. (2004) *Badiou: Theoretical Writings*, Continuum: London.

Barkun, M. (2010) The "New World Order" and American Exceptionalism, in

Barringer, T. and Wilton, A. (2002) *American Sublime: Landscape Painting in the United States 1820-1880*, Princeton University Press: New Jersey.

Barthes, R. (1957/2000) *Mythologies*, Random House: London.

Baudrillard, J. (1983) *Simulations*, Semiotext[e]: New York.

Beery, J. (2012) State, capital and spaceships: A terrestrial geography of space tourism, *Geoforum*, 43(1), 25-34.

Bell, D. and Parker, M. (2009b) Introduction: making space in Bell, D. and Parker, M. (Eds) *Space Travel & Culture: From Apollo to Space Tourism*, Wiley-Blackwell: Oxford, pp. 1-5.

Bell, D. and Parker, M. (Eds) (2009a) *Space Travel & Culture: From Apollo to Space Tourism*, Wiley-Blackwell: Oxford.

Beniger, J. (1986) *The Control Revolution: Technological and Economics Origins of the Information Society,* Harvard University Press: Cambridge, Mass.

Benjamin, M. (2003) *Rocket Dreams. How the Space Age Shaped our Vision of a World Beyond*, Chatto and Windus: London.

Benjamin, W. (1970/1999) Thesis on the Philosophy of History in Arendt, H. (Ed.) *Illuminations*, Pimlico: New York, pp. 245-255.

Benson, C. D. and Faherty, W. B. (1978). *Moonport: A History of Apollo Launch Facilities and Operations* (NASA SP4204), NASA History Office: Washington DC.

Beschloss, M.R. (1997) Kennedy and the Decision to Go to the Moon, in Launius, R, D. and McCurdy, H.E. (Eds) *Spaceflight and the Myth of Presidential Leadership*, University of Illinois Press: Champaign, IL, pp. 50-67.

Beynon, J. (2002) *Masculinities and Culture*, Open University: Buckingham.

Billig, M. (1995) *Banal Nationalism*, Sage: London.

Bingham, C. (2000) Oral history interview for Johnson Space Center Oral History Project [Online] Available: www.nasa.gov/jsc.htm [15/5/07].

Bizony, M. (2006) *The Man who Ran the Moon*, Thunder's Mouth Press: New York.

Boime, A. (1991) *The Magisterial Gaze: Manifest Destiny and American Landscape Painting, circa 1830-1865*, Smithsonian Institution Press: Washington D.C.

Bonta, M. and Protevi, J. (2006). *Deleuze and Geophilosophy: A Guide and Glossary*, Edinburgh University Press: Edinburgh.

Booker, M. K. (2001) *Monsters, Mushroom Clouds and the Cold War*, Greenwood Press: Connecticut.

Bostick, J. (2000) Oral history interview for Johnson Space Center Oral History Project [Online] Available: www.nasa.gov/jsc.htm [15/5/07]

Broad, W, J. (1986) Rain Clouds Hover Near the Site of Shuttle's Flight Set for Today, *New York Times*. January 26th, p. 16.

Bromberg, J.L. (1999) *NASA and the Space Industry*, John Hopkins University Press: Baltimore.

Bryan, B. and Dixon, C. (1979) *The National Air and Space Museum*, Harry N. Abrahams: New York.

Bryman, A. (2004) *The Disneyization of Society*, Sage: London.

Buchanan (2005) *Deleuze and Space*, Edinburgh University Press: Edinburgh.

Burrows, W. E. (1998) *This New Ocean*, The Modern Library: New York.

Cabbage, M. and Harwood, W. (2004) *Comm Check ...: The final flight of Shuttle Columbia*, Free Press: New York.

Cadbury, D. (2005) *Space Race*, Harper Collins: London.

Carrigan, T., Connell, R. and Lee, J. (1985) Towards a new sociology of masculinity. *Theory and Society*, 14(5), pp. 551-604

Carter, F. (1936) *Phoenix: The Posthumous Papers of D.H. Lawrence*, Viking Press: New York.

Casey, E. (2002) *Representing Place: Landscape Painting and Maps*, University of Minnesota Press: Minneapolis, MN

Cashford, J. (2003) *The Moon: Myth and Image*, Four Walls Eight Windows: New York.

Chaikin, A. (1998) *A Man on the Moon*, Penguin: London.

Clancey, W, J. (2012) *Working on Mars: Voyages of Scientific Discovery with the Mars Exploration Rovers*, MIT Press: Cambridge, Mass.

Clegg, S. and Kornberger, M. (2004) Bringing Space Back In: Organizing the Generative Building, *Organization Studies*, 25(7), 1095-1114.

Clegg, S., Courpasson, D. and Phillips, N. (2006) *Power and Organizations*, Sage: London.

Collis, C. (2009) The Geostationary Orbit: a Critical Legacy of Space's Most Valuable Real Estate in Bell, D. and Parker, M. (Eds) *Space Travel & Culture: From Apollo to Space Tourism*, Wiley-Blackwell: Oxford, pp. 47-65

Congress Report (1968) Report number 956—Apollo 204 report together with additional views' 90th Congress, 2nd Session, January 30th, 1968, NASA History Office: Washington DC

Connell, R.W. (1995) *Masculinities*, University of California Press: Berkeley, CA.

Connolly, W. (2002) *Neuropolitics: Thinking, Culture and Speed*, University of Minnesota Press: Minneapolis, MN.

Cook, R. C. (2006) *Challenger Revealed: An Insider's Account of How the Reagan Administration Caused the Greatest Tragedy of the Space Age*, Thunder Mouth Press: New York City, NY.

Cooper, J. F. (1832/2000) *The Pioneers: Oxford World Classics Edition*, Oxford University Press: Oxford.

Corona, V, P. and Godart, F, C. (2010) Network-Domains in Combat and Fashion Organizations, *Organization*, 17(2), 283-304.

Cosgrove, D. (2001) *Apollo's Eye*, John Hopkins University Press: Baltimore, MD.

Cowan, R. S (1976) Two Washes in the Morning and a Bridge Party at Night: The American Housewife Between the Warts, *Women's Studies* 3(2), 147-72

Crane, S, A. (1997) Memory, Distortion, and History in the Museum, *History and Theory*, 36(4), 44-63

Crawford, J. (1997) *James Fenimore Cooper: His Country and His Art (No. 11)* [Online]Available: http://external.oneonta.edu/cooper/articles/suny/1997suny-pikus.html [12/5/12].

Crouch, T. D. (1997) Risky business: some thoughts on controversial exhibitions, *Museum International*, 49(3), 8-13.

Crowther, P. (1989) *The Kantian Sublime: From Morality to Art*, Oxford: Clarendon Press.

Dale, K. and Burrell, K. (2008) *The Spaces of Organisation and the Organisation of Space: Power, Identity and Materiality at Work*, Palgrave Macmillan: Basingstoke.

Dee, G. (1986) Soviets Link Accident to SDI, *Washington Post*, January 31st, p. A12

DeLanda, M. (2006) *A New Philosophy for Society*, Continuum: London.

Deleuze, G. and Guattari, F. (1988) *A Thousand Plateaus*, Continuum: London.

Deleuze, G. and Guattari, F. (1994) *What is Philosophy?*, Columbia: New York.

Derrida, J. (2002) Declarations of Independence, in Rottenberg, E. (Ed.) *Negotiations: Interventions and Interviews, 1971-2001*, Stanford University Press: Stanford, CA, pp. 46-54.

Dickens, P and Ormrod, J, S. (2007) *Cosmic Society: Towards Sociology of the Universe*, Routledge: London.

Dickson, P. (2001) *Sputnik: The Shock of the Century*, Berkeley Books: New York.

Dijkink, G. (2006) When Geopolitics and Religion Fuse: A Historical Perspective, *Geopolitics*, 11(2), 192-208

Dittmer, J. (2007) Colonialism and Place Creation in Mars Pathfinder Media Coverage, *The Geographical Review*, 97(1), 112-30.

Dittmer, J. and Strum, T. (Eds) (2010) *Mapping the End Times: American Evangelical Geopolitics and Apocalyptic Visions*, Ashgate: London, pp. 119-132.

Dittmer, J. and Strum, T. (Eds) (2010) *Mapping the End Times: American Evangelical Geopolitics and Apocalyptic Visions*, Ashgate: Farnham.

Dolman, E. (1999) Geostrategy in the Space Age: an astropolitical analysis, in Gray C. and Sloan G. (Eds), *Geopolitics, Geography and Strategy*, Frank Cass: London, pp. 83-105

Dolman, E. (2001) *Astropolitik: Classical Geopolitics in the Space Age*, Frank Cass: London.

Duke, C. (1990) *Moonwalker*, Thomas Nelson Inc: Nashville, TN.

Dunnett, O. (2012) Patrick Moore, Arthur C. Clarke and 'British Outer Space' in the mid 20th century, *Cultural Geographies*, 19(4), 505-22.

Elden, S (2013) Secure the volume: Vertical geopolitics and the depth of power, *Political Geography*, 34, 35-51.

Eliot, T, S. (1944/2001) Four Quartets, Faber and Faber: London.

Emerson, R. W. (1836/2003). *Nature*, Penguin: London.

Feenberg, A. (1996) From Essentialism to Constructivism: Philosophy of Technology at the Crossroads [Online] Available: www.rohan.sdsu.edu/faculty/feenberg/talk4.html [15/7/07]

Feenberg, A. (1999) *Questioning Technology*, Routledge: London.

Fesunenko, I. (1986) World Today program Moscow Television Service (in Russian) 11.45 GMT 31st January 31st, Available in English: NASA History Office, Washington DC.

Fletcher, J. (1986) Remarks prepared for delivery: Aerospace Industries Association of America Board of Governors Meeting, Williamsburg, Virginia. May 22nd 1986, NASA press release #86-66, Available: NASA History Office, Washington D.C.

Ford, G. (1976) *Remarks at the Dedication Ceremony for the National Air and Space Museum at the Smithsonian Institute*, July 1st, Gerald R. Ford Presidential Library, Ann Arbor, MI.

GAP (1990) Challenger advertisement, *Recreation News*, November, p. 21

Geppert, A, C, T. (2012a) *Imagining Outer Space: European Astroculture in the Twentieth Century*, Palgrave Macmillan: Basingstoke.

Geppert, A, C, T. (2012b) European Astrofuturism, Cosmic Provincialism: Historicizing the Space Age. in Geppert, A, C, T. (2012a) *Imagining Outer Space: European Astroculture in the Twentieth Century*, Palgrave Macmillan: Basingstoke. pp. 3-26

Gieryn, T. F. (1998) Balancing acts: science, Enola Gay and History Wars at the Smithsonian, in Macdonald, S. (Ed) *The Politics of Display: Museums, Science, Culture*, Routledge: London, pp. 197-228

Godfrey, R. Liley, S. and Brewis, J. (2012) Biceps, Bitches and Borgs: Reading *Jarhead*'s Representation of the Construction of the (Masculine) Military Body, *Organization Studies*, 33(4), 541-62

Graham, S. (2004) Vertical Geopolitics: Baghdad and After, *Antipode*, 36(1), 12-23

Graham, S. and Hewitt, L. (2013) Getting off the Ground: on the Politics of Urban Verticality, *Progress in Human Geography*, 37(1), 72-92.

Gray, T. (1998) A Brief History of Animals in Space. NASA History Office: Washington DC. Available: www.history.nasa.gov/apollo204 [12/4/06]. Also available from: http://history.nasa.gov/printFriendly/animals.htm. [20/5/07].

Gregory, D. (2011) From a View to a Kill: Drones and Late Modern War, *Theory, Culture & Society*, 28(7-8), 188-215.

Grey, C. (2003) The Real World of Enron's Auditors, *Organization*, 10(3), 572-576.

Grey, C. (2009) Security Studies and Organization Studies: Parallels and Possibilities, *Organization*, 16(2), 303-16.

Grosz, E. (1989) *Sexual Subversions: Three French Feminists*, Unwin and Hyman, London.

Gutterman (2001) Postmodernism and the Interrogation of Masculinity, in Whitehead, S and Barret, F. (Eds) (2001) *Masculinities Reader*, Polity Press: Cambridge, pp. 56-71.

Hamacher, D, W. (2011) Meteoritics and Cosmology among the Aboriginal Cultures of Central Australia, *Journal of Cosmology*, 13, 3743-53.

Haraway, D, J. (1991) *Simians, Cyborgs, and Women: The Reinvention of Nature*, Free Association Books: London.

Haraway, D, J. (2004) *The Haraway Reader*, Routledge: London.

Hardy, D. (2002) Faces of the Moon, *Spaceflight*, 45, 103-5.

Hendershot, C. (1999) *Paranoia, The Bomb, and 1950s Science Fiction Films*, Bowling Green State: OH.

Henry, H. and Taylor, A. (2009) Re-thinking Apollo: Envisioning Environmentalism in Space, in Bell, D. and Parker, M. (Eds) *Space Travel & Culture: From Apollo to Space Tourism*, Wiley-Blackwell: Oxford, pp. 190-203

Hersch, M, H. (2009) Checklist: The Secret life of Apollo's 'fourth crewmember' in Bell, D. and Parker, M. (Eds) *Space Travel & Culture: From Apollo to Space Tourism*, Wiley-Blackwell: Oxford, pp. 6-24

Hersch, M, H. (2012) *Inventing the American Astronaut*, Palgrave Macmillan: Basingstoke.

Hess, D. J. (1997) *Science Studies: An Advanced Introduction*, New York: New York University Press.

Hetherington, K. (2006), Museum. *Theory, Culture and Society*, 23 (2-3), 597-603

Hine, T. (1992) A time machine through space, *The Philadelphia Inquirer*, 20th January, pp. 22-3

Hines, W. (1967) Three Apollo astronauts killed by flash fire in craft on pad, *The evening Star* [Washington DC], January 28th

Hjalmarson, B. (1999) *Artful Players: Artistic Life in Early San Francisco*, Balcony Press: California.

Hodgson, D. and Cicmil, S. (Eds) (2006) *Making Projects Critical*, Palgrave Macmillan: Basingstoke.

Hoff, J. (1997) The Presidency, Congress, and the Deceleration of the U.S. Space Program in the 1970s, in Launius, R, D. and McCurdy, H, E. (Eds) *Spaceflight and the Myth of Presidential Leadership*, University of Illinois Press: Champaign, IL, pp. 93-132

Hoffe, O. (1994) *Immanuel Kant*, State University of New York Press: Albany, NY.

Howat, J, K (1987) *American Paradise: The World of the Hudson River School*, The Metropolitan Museum of Art: New York City, NY.

Huntoon, C. (2002) Oral history interview for Johnson Space Center Oral History Project [Online]. Available: www.nasa.gov/jsc.htm [15/5/07].

Huyssen, A. (1995) *Twilight Memories: Marking Time in a Culture of Amnesia*, Routledge: London.

Isikoff, M. (1986) Thiokol Was Seeking New Contract When Officials Approved Launch, *Washington Post*, February 27th, p. A15.

Johns, J. (2007) Nature and the American Identity [Online]. Available: http://xroads.virginia.edu/~cap/NATURE/cap2.html [20/2/07].

Johnson, J. (1986) NASA's Decaying Public Image, *Washington Post*, February 12th p. A2.

Johnson, S. B. (2002) *The Secrets of Apollo: Systems Management in American and European Space Programs*, John Hopkins University Press: Baltimore, MD.

Jones, H. (1996) Landscape Painters of Northern California 1870-1930, Traditional Fine Arts Organization [Online]. Available: http://www.tfaoi.com/aa/4aa/4aa320.htm [12/9/06].

Jones, L. and Sage, D. (2010) New Directions in Critical Geopolitics, *Geojournal*, 75(4), 315-25.

Jorden, W. J. (1957) Soviet Goes all out on Satellite Bonaza, *New York Times*, 6th October 1957.

Jorgensen, D. (2009) Middle America, the Moon, the Sublime and the Uncanny. in Bell, D. and Parker, M. (Eds) *Space Travel & Culture: From Apollo to Space Tourism*, Wiley-Blackwell: Oxford, pp. 178-189.

Kaldor, M. (2003) American Power: from 'Compellance' to 'Cosmopolitanism,' *International Affairs* 79(1), 1-22.

Karegeannes, C., Wells, H. T. and Whiteley, S, H. (1976) Origins of NASA Names, (SP-4402), NASA History Office: Washington DC.

Kelly, T. (2000) Oral history interview for Johnson Space Center Oral History Project [Online]. Available: www.nasa.gov/jsc.htm [15/5/07]

Kennedy, J, F (1962) Address at Rice University on the space effort, September 12, 1962 [Online]. Available: http://www.rice.edu/fondren/woodson/speech.html [2/3/06].

Kennedy, J, F. (1961b) Special Message to the Congress on Urgent National Needs, speech given before a Joint Session of Congress May 25th, 1961, John F. Kennedy Presidential Library and Museum: Boston, MA.

Kennedy, J, F. (1961c) Inaugural address by John F. Kennedy, January 20th John F. Kennedy Presidential Library and Museum: Boston, MA.

Kennedy, J. (1961a). Memorandum for Vice President, April 20th, 1961, John F. Kennedy Presidential Library and Museum: Boston: MA.

Kennedy, J. (1963) Conversation between Kennedy and James Webb, September 18th, 1963, Kennedy Presidential Library and Museum: Boston, MA.

Kerwin, J, P. (1986) Report from Joseph P. Kerwin to Richard H. Truly on Challenger Disaster, July 28, 1986 [Online]. Available: http://history.nasa.gov/kerwin.html [3/2/13].

Kessler, E, A. (2012) *Picturing the Cosmos*, University of Minnesota Press: Minneapolis, MN.

Kevles, B. H. (2003) *Almost Heaven the Story of Women in Space*, Basic Books: Jackson TN.

Kiernan, K. (1986) Suddenly, the Celebration Stopped, *Washington Post*, January 29th, p. A1.

Kinsey, J, L. (1992) *Thomas Moran and the Surveying of the American West*, Smithsonian Institution Press: Washington DC.

Klerkx, G. (2004) *Lost in Space: the Fall of NASA and the Dream of a New Space Age*, Pantheon: New York City, NY.

Kornhauser, E, M. (2003) *Masterworks from the Wadsworth Atheneum Museum of Art*, Yale University Press: Yale, CT.

Kristeva, J. (1982) *The Powers of Horror: an Essay on Abjection*, Columbia University Press: New York City, NY.

KSC (2005a) Thematic Development Plan, media release, July 6th 2005, Kennedy Space Center Visitor Complex, FL 32899.

KSC (2005b) Thematic Development Plan: Shuttle Launch Experience Project Team, media release, July 6th 2005, Kennedy Space Center Visitor Complex, FL 32899.

KSC (2005c) Information from Delaware North Companies on TDP, media release, July 6th 2005, DNC Parks and Resorts at KSC, Inc., Kennedy Space Center Visitor Complex, FL 32899.

KSC (2006a) Guided Tours promotional leaflet, Kennedy Space Center Visitor Complex, FL 32899.

KSC (2006b) *Official Tour book*, Kennedy Space Center Visitor Complex, FL 32899.

KSC (2006c) Destination Kennedy Space Center media release, January 2006, Kennedy Space Center Visitor Complex, FL 32899.

KSC (2006d) Kennedy Space Center: Featured Facts, media release, January 2006, Kennedy Space Center Visitor Complex, FL 32899.

KSC (2006e) Kennedy Space Center Visitor Complex: Fact Sheet, media release January 2006, Kennedy Space Center Visitor Complex, FL 32899.

KSC (2014) Kennedy Space Centre 1966, John Proctor [Online]. Available: http://jonproctor.net/1966-kennedy-space-center/ [1/2/14].

Lane, M. (2011) *Geographies of Mars: Seeing and Knowing the Red Planet*, University of Chicago Press: Chicago, IL.

Latour, B. (1987) *Science in Action: How to Follow Scientists and Engineers Through Society*, Milton Keynes: Open University Press.

Latour, B. (2005) *Reassembling the Social: an Introduction to Actor-Network Theory*, Oxford University Press: Oxford.

Latour, B. (2013) *An Inquiry into Modes of Existence*, Harvard University Press: Cambridge, MA.

Launius, R, D. (2006) Assessing the legacy of the Space Shuttle, *Space Policy*, 22(4), 226-34.

Launius, R, D. and McCurdy, H, E. (1997) Epilogue: Beyond NASA Exceptionalism, in Launius, R, D. and McCurdy, H, E. (Eds) *Spaceflight and the Myth of Presidential Leadership*, University of Illinois Press: Champaign, IL, pp.221-250.

Launius, R. D. (2005) Sputnik and the Origins of the Space Age [Online]. Available: http://history.nasa.gov/sputnik/sputorig.html [5/5/06].

Law, J. (2006) *After Method: Mess in Social Science Research*, Routledge: London.

Lawrence, W. H. (1957) President Voices Concern on U.S. Missiles Program, *New York Times,* 10th October.

Lefevbre, H. (1991) *The Production of Space*, Wiley-Blackwell: Oxford.

Lester, D, F. and Robinson, M. (2009) Visions of exploration, *Space Policy*, 25(4), 236-43.

Ley, W. (1950) *The Conquest of Space*, Sidgwick and Jackson Limited: London.

Lindauer, M. (2006) The Critical Museum Visitor, in Marstine, A. (Ed.) *New Museum Theory and Practice: an introduction*, Blackwell: London, pp. 203-25.

Livingstone, D. N. (1993) *The Geographical Tradition: Episodes in the History of a Contested Enterprise*, Blackwell: London.

Logsdon, J. (1969) *Decision to Go to the Moon*, University of Chicago Press: Chicago, IL.

Logsdon, J. (1994) Apollo Exemplifies What Not to Do, *Space News*. July, 18-24

Logsdon, J. (1995) The Evolution of U.S. Space Policy, in Logsdon, J., Lear, L, J. Warren-Findley, J. Williamson, R, A. and Day, D, W. (Eds) *NASA and the Exploration of Space: volume 1* (SP-4218), NASA History Office: Washington DC. pp. 377-93.

Logsdon, J. (1997) National Leadership and Presidential Power, in Launius, R, D. and McCurdy, H, E. (Eds.) *Spaceflight and the Myth of Presidential Leadership*, University of Illinois Press: Champaign, IL, pp. 204-19.

Logsdon, J., Lear, L, J. Warren-Findley, J. Williamson, R, A. and Day, D, W. (Eds.) (1995) *NASA and the Exploration of Space: volume 1* (SP-4218), NASA History Office: Washington DC.

Macauley, W, M. (2012) Crafting the Future: Envisioning Space Exploration in Post-war Britain, *History and Technology: an International Journal*, 28(3), 281-309

MacDonald, F. (2007) Anti-Astropolitik: Outer Space and the Orbit of Geography *Progress in Human Geography*, 31(5), 592-615.

MacDonald, F. (2008) Space and the Atom: On the Popular Geopolitics of Cold War Rocketry, *Geopolitics*, 13(4), 611-34.

Maloney, J. (1977) Fatal Apollo fire 10 years ago, *Washington Post*, January 27th.

Marsella, A, J. (2011) The United States of America: "A Culture of War," *International Journal of Intercultural Relations*, 35(6), 714-28.

Martstine, A. (2006) *New Museum Theory and Practice: An Introduction*, Blackwell: London.

Massumi, B. (2002) *Parables for the Virtual: Movement, Sensation and Affect*, Duke University Press: Durham, NC.

Maxwell, W, D. (1967) A second space test tragedy, *Chicago Tribune*, Feb 1st, p.4.

McConnell, M. (1987) *Challenger, a Major Malfunction—A True Story of Politics, Greed and the Wrong Stuff*, Doubleday and Company: New York City, NY.

McCurdy, H, E. (1993) *Inside NASA: High Technology and Organizational Chance in the U.S. Space Program*, John Hopkins University Press: Baltimore, MD.

McCurdy, H, E. (1997) *Space and the American Imagination*, Smithsonian Institution Press: Washington DC.

McCurdy, H, E. (2013) Learning from History: Low-cost project innovation in the U.S. National Aeronautics and Space Administration, *International Journal of Project Management*, 31(5), 705-11.

McDougall, W. A. (1985) *A Political History of the Space Age*, John Hopkins University Press: Baltimore, MD.

McDougall, W. A. (1997) *Promised Land, Crusader State: an American encounter with the world since 1776*, Mariner Books: Boston, MA.

McGaw, J. (1987) *Most Wonderful Machine: Mechanization and Social Change in Berkshire Paper Making, 1801-1895*, Princeton University Press: New Jersey, NJ.

McNamara, R. and Webb, J. (1961) Recommendations for Our National Space Program: Changes, Policies and Goals. May 8th 1961, in Logsdon, J., Lear, L, J. Warren-Findley, J. Williamson, R, A. and Day, D, W. (Eds.) *NASA and the Exploration of Space: volume 1* (SP-4218), NASA History Office: Washington DC. pp. 439-52.

MDS (1986) Vladimir Paska Commentary, Moscow Domestic Service [in Russian] 13.00 GMT 16th August, Available in English: NASA History Office: Washington DC.

Meslay, O. (2004) *J.M.W. Turner: the man who set painting on fire*, Thames and Hudson: London.

Miller, R. (2007) Biography of Chesley Bonestell [Online]. Available: www. bonestellart.com [12/2/07]

Miller, R. and Durant III, F. (2001) *The Art of Chesley Bonestell*, Paper Tiger: London.

Musk, E. (2012) SpaceX Billionaire Elon Musk On The Business And Future Of Space Travel, *Forbes*, 23rd April 2012 [Online]. Available: http://www. forbes.com/sites/alexknapp/2012/04/23/spacexs-elon-musk-on-the-business-and-future-of-space-travel/ [12/10/13]

NASA (1965) Summary Report: Future Program Task Group, in Logsdon, J., Lear, L, J. Warren-Findley, J. Williamson, R, A. and Day, D, W. (Eds.) *NASA and the Exploration of Space: volume 1* (SP-4218), NASA History Office: Washington DC, pp. 473-89.

NASA (1967) Report of Apollo 204 Review Board [Online]

NASA (1988) MCC Status Report #17 STS-26: Crew Tribute to Challenger Crew October 2nd, NASA History Office: Washington DC.

NASA (2013) Space Shuttle Basics: Launch [Online]. Available: http:// spaceflight.nasa.gov/shuttle/reference/basics/launch.html [2/2/14]

NASA (2014) NASA Budget Information, FY2012 [Online]. Available: http:// www.nasa.gov/news/budget/2012.html [10/2/14]

NASA History Office (2007) NASA History Office, NASA Headquarters, Washington, DC.

NASM (2013) Facts and Figures—Smithsonian National Air and Space Museum [Online] Available: http://airandspace.si.edu/about/facts.cfm? [2/8/13]

New York Times (1986a) At Mission Control, Silence and Grief, *New York Times*, 28th January.

New York Times (1986b) Gorbachev Expresses His Condolences, *New York Times*, January 30th, p. A16

New York Times (1986c) Journalist Says NASA's Reticence Forced Them to Gather Data Elsewhere, *New York Times,* February 9th.

Newsweek (1969) Opinion Poll on Space Travel, October 6th.

Nietzsche, F. (1885/1990) *Beyond Good Evil*, Penguin: London.

NIUPOL (1987) Report on The Net Effect of the Challenger Accident: Impact of the Challenger on Public Attitudes towards the space program, January 25th, National Science Foundation, Arlington, VA.

Noble, D, W. (2002) *Death of a Nation: American Culture and the End of Exceptionalism*, University of Minnesota Press: Minneapolis, MN.

Novak, B. (1995) *Nature and Culture: American Landscape and Painting, 1825-1875*, Oxford University Press: Oxford.

Nye, D, E. (1996) *American Technological Sublime*, MIT Press: Cambridge, MA.

Ó Tuathail, G. (1996*) Critical Geopolitics*, University of Minnesota Press: Minneapolis, MN.

Ó Tuathail, G. (1999) Understanding critical geopolitics: Geopolitics and Risk Society, *Journal of Strategic Studies*, 22(2-3), 107-24

Ó Tuathail, G. (2000) Dis/placing the Geo-Politics which One Cannot Not Want, *Political Geography*, 19, 385-96.

Ó Tuathail, G.and Dalby, S. (1998) Introduction, in Ó Tuathail, G. and Salby, S. (Eds) *Rethinking Geopolitics*, London, Routledge, pp. 1-15.

O' Toole, T. (1967a) NASA Accused of Covering Up Troubles, *Washington Post*, May 11th, p.A9.

O'Brien, C, C. (1988) *God Land: Reflections on Religion and Nationalism*, Harvard University Press: Cambridge, MA.

O'Toole, T. (1967b) House Panel Clouds Outlook for NASA, *Washington Post*. May 17th, p. A1.

Obama, B. (2009) Obama hails Apollo 11 astronauts, BBC News, 20th July 2009 [Online]. Available: http://news.bbc.co.uk/1/hi/sci/tech/8160209.stm [12/9/13]

Obama, B. (2010) President Obama's Speech at Kennedy, 15th April 2010 [Online]. Available: http://www.nasa.gov/about/obama_ksc_pod.html [15/10/13]

Oberg, J. (2007) *Space Power Theory* [Online]. Available: http://space.au.af.mil/books/oberg/ [5/7/07]

Olsson, G. (2007) *Abysmal: A Critique of Cartographic Reason*, University of Chicago Press: Chicago, IL.

Pacey, A. (1999) *Meaning in Technology*, MIT Press: Cambridge, MA.

Parker, M. (2009a) Space Age Management, *Management and Organizational History*, 4(3), 317-332.

Parker, M. (2009b) Capitalists in Space, in Bell, D. and Parker, M. (Eds) *Space Travel & Culture: From Apollo to Space Tourism*, Wiley-Blackwell: Oxford. pp. 83-97

Parker, M. (2013) Containerisation: Moving Things and Boxing Ideas, *Mobilities*, 8(3), 368-87.

Parks, L. (2005) *Cultures in Orbit: Satellites and the Televisual*, Duke University Press: Durham, NC.

Peckham, R. S. (2003) Introduction: the Politics of Heritage and Public Culture, in Peckham, R, S. (Ed.), *Rethinking Heritage*, I.B Tauris: London. pp. 1-16

Peltakian, D. (2006) *Albert Bierstadt: Landscape Painter of the American West* [Online]. Available: http://www.sullivangoss.com/albert_Bierstadt/#Biography [12/3/06]

Peoples, C. (2009) *Justifying Ballistic Missile Defence: Technology, Security and Culture*, Cambridge University Press: Cambridge.

Plumb, R. K. (1957) New Space Conquests Can Now be Foreseen, *New York Times* 6th October 1957.

Powell, J. (1986) Correspondence: Recollections, *Spaceflight magazine*, 28(April)

Preziosi, D. (2003) The Museum of What You Shall Have Been, in Peckham, R, S. (Ed) *Rethinking Heritage*, I.B Tauris: London, pp. 169-81

Prosise, T, O. (1998) The collective memory of the atomic bombings misrecognized as objective history: The case of the public opposition to the national air and space museum's atom bomb exhibit, *Western Journal of Communication*, 62(3), 316-47.

Reagan, R. (1982) Remarks at Edwards Air Force Base, California, on Completion of the Fourth Mission of the Space Shuttle Columbia, July 4th [Online]. Available: http://www.reagan.utexas.edu/archives/speeches/1982/70482a. htm [2/2/14]

Reagan, R. (1986) Transcript of President Reagan's News Conference, *Washington Post* June 12th, p. A18

Redfield, P. (2002) The Half-Life of Empires in Outer Space, *Social Studies of Science*, 32(5-6), 791-825.

Reistrup, J. V. (1967a) Senate Report Blasts NASA as Negligent, *Washington Post*, 22nd December, p.A1

Reistrup, J. V. (1967b) NASA Builds Quick Apollo Escape Hatch, *Washington Post*, 28th February, p. A3

Rensverg, B. (1986) Fire Engulfs Ship with 7 aboard soon after liftoff, *Washington Post*, 29th January, p.A1

Ricard, S. (1999) The 'Manifest Destiny' of the United States in the 19th Century, Didier Erudition, CNED: Paris.

Rice, W (2004) Oral history interview for Johnson Space Center Oral History Project [Online]. Available: www.nasa.gov/jsc.htm [15/15/07].

Richardson, R. S. (1948) Rocket Blitz from the Moon. *Collier's,* 23rd October 1948, pp. 24-25, 44-46.

Rogathevski, A. (2011) Space Exploration in Russian and Western Popular Culture: Wishful Thinking, Conspiracy Theories and Other Related Issues, in Maurer, E. Richards, J, Rüthers, M. and Scheide, C. (Eds) *Soviet Space Culture: Cosmic Enthusiasm in Socialist Societies*, Palgrave Macmillan: Basingstoke, pp. 251-65.

Roland, A. (1985) The Shuttle: triumph or turkey? *Discover*, November, 14-24.

Romanowski, D.A. (2002) *Official Guide to the Smithsonian National Air and Space Museum*, Smithsonian Institution Press: Washington DC.

Roper, M. and Tosh, J. (Eds) (1991) *Manful Assertions: Masculinities in Britain since 1800*, Routledge: London

Sagan, C. (1994) *Pale Blue Dot: A Vision of the Human Future in Space*, Random House: New York City, NY.

Sage, D. (2009) Giant leaps and forgotten steps: NASA and the performance of gender, in Bell, D. and Parker, M. (Eds) *Space Travel & Culture: From Apollo to Space Tourism*, Wiley-Blackwell: Oxford, pp. 146-63.

Sawyer, K. (1986) The Horror Dawned Slowly, *Washington Post*, 29th January, p. A1

Sehlstedt, R. (1969) Reports pending on Apollo fire, *Baltimore Sun*, 27th January. p. A4,

Serres, M. and Latour, B. (1995) *Conversations on Science, Culture, and Time: Michel Serres Interviewed by Bruno Latour*, University of Michigan Press: Ann Arbor, MI.

Sharp, J. (2000) *Condensing The Cold War: Reader's Digest and American Identity* Minneapolis: University of Minnesota Press, MN.

Shaw, P. (2006) *The Sublime*, Routledge: London.

Shayler, D. J. and Moule, I. A. (2003) *Women in Space—Following Valentina*, Springer: New York City, NY.

Singer, R. (2010) Neoliberal Style, the American Re-Generation, and Ecological Jeremiad in Thomas Friedman's "Code Green," *Environmental Communication: A Journal of Nature and Culture*, 4(2), 135-51.

Smithsonian (2013) History of the Smithsonian, Smithsonian Institution [Online]. Available: http://www.si.edu/About/History [5/9/13]

Smolkin-Rothrock, V. (2011) The Contested Skies: The Battle of Science and Religion, in the Soviet Planetarium in Maurer, E. Richards, J, Rüthers, M.

and Scheide, C. (Eds) *Soviet Space Culture: Cosmic Enthusiasm in Socialist Societies*, Palgrave Macmillan: Basingstoke, pp. 57-78

Steinberg, P, E. (2001) *The Social Construction of the Ocean*, Cambridge University Press: Cambridge.

Stephanson, A. (1995) *Manifest Destiny: American Expansion and the Empire of Right*, Hill and Wang: New York City, NY.

STG (1969) *Space Task Group, The Post-Apollo Space Program: Directions for the Future, September 1969*, in Logsdon, J., Lear, L, J. Warren-Findley, J. Williamson, R, A. and Day, D, W. (Eds.) (1995) *NASA and the Exploration of Space: volume 1* (SP-4218), NASA History Office: Washington DC, pp. 522-543.

Stich, S. (2003) Email from Steve Stich to Columbia Crew, January 23rd 2003 [Online]. Available: http://www.nasa.gov/columbia/home/index.html [12/2/06]

Stine, H. (1996) *Halfway to Anywhere*, M. Evans and Company: New York City, NY.

Strum, T. (2010) Imagining Apocalyptic Geopolitics: American Evangelical Citationality of Evil Others, in Dittmer, J. and Strum, T. (Eds) (2010) *Mapping the End Times: American Evangelical Geopolitics and Apocalyptic Visions*, Ashgate: London, pp.133-156

Strum, T. (2013) The future of religious geopolitics: towards a research and theory agenda, *Area*, 45(2) 134-40.

Strum, T. and Dittmer, J. (2010) Introduction: Mapping the End Times, in Dittmer, J. and Strum, T. (Eds) (2010) *Mapping the End Times: American Evangelical Geopolitics and Apocalyptic Visions*, Ashgate: Farnham, pp.1-26

Sumida, J. (1999) Alfred Thayer Mahan, Geopolitician, in *Geopolitics: Geography and Strategy*, Gray, C. and H. Sloan, G. (Eds.) Frank Cass: London, pp. 39-62

Suro, R. (1986) 'Deep Sorrow in My Soul,' Pope Says, *New York Times*, 29th January.

Tapscott, M. (1986) President expected to phase out commercial payloads on shuttle, *The Washington Times*, June 18th, p. 10A.

Tass (1986) Moscow Tass Agency, English 18.15 GMT 31st January, NASA History Office: Washington DC.

Taylor, F, W. (1911) *The Principles of Scientific Management*, Harper & brothers: New York City, NY.

Taylor, R. (Ed.) (1972) *The Turner Thesis: Concerning the Role of the Frontier in American History*, D.C. Heath and Company: Washington DC.

Terr, L. (1992) *Too scared to cry: psychic trauma and childhood*, Basic Books: New York City, NY.

TFAOI (2006) *Art and the Empire City: New York, 1825-1861 September 19, 2000—January 7, 2001* [Online]. Available: http://www.tfaoi.com/aa/2aa/2aa253.htm [12/9/06]

Thoreau, H. D. (1854/1989) Walden, Princeton University Press: New Jersey, NJ.

Thrift, N. (2000) It's the Little Things, in Dodds K, and Atkinson D. (Eds) *Geopolitical Traditions: a Century of Geopolitical Thought*, Routledge: London, pp. 380-7.

Thrift, N. (2008) *Non-representational Theory: Space, politics, affect*, Routledge: London.

Tsiolkovsky, K. E. (1911) [Letter] Kaluga, Russia.

Tuan, Y-F. (1993) *Passing Strange and Wonderful: Aesthetics, Nature, and Culture*, Island Press, Shearwater Books: Washington, DC.

UN (2008) United Nations Outer Space Treaty, United Nations Office for Outer Space Affairs [Online]. Available: http://www.unoosa.org/oosa/en/SpaceLaw/gares/html/gares_21_2222.html. [12/2/08]

Vaughan D. (1996) *The Challenger Launch Decision: Technology, Deviance and Risk* University of Chicago Press: Chicago, IL.

Virilio, P. (1999) *Politics of the Very Worst*, MIT Press: Cambridge, MA.

Vobejda, N. (1986) Schools Here and Across Nation Shaken but Still Support Program *Washington Post*, January 29th, p.A1.

Wallace, I. (2006) Territory, Typography, Theology: Geopolitics and the Christian Scriptures, *Geopolitics*, 11(2), 209-30.

Wallace, M. (1996) *Mickey Mouse History and other Essays on American Memory*. Temple University Press: Philadelphia, PA.

Wallis, B. (1994) Selling Nations. in *Museum Culture: Histories, Discourses and Spectacles*, in Sherman, D, J. and Rogoff, I. (Eds) University of Minnesota Press: Minneapolis, MN, pp. 265-81

Ward, B. (1966) *Spaceship Earth*, Columbia University Press: New York City, NY.

Ward, B. (2006) *From Nazis to NASA: The Life of Wernher von Braun*, The History Press: Stroud.

Washington Post (1987) Journalists' Ad Angers NASA: Challenger Disaster Used to Depict Benefits of Free Press, *Washington Post*, 29th May, p. A23.

Washington Post (1996) The Picture of Disaster: Seeing Politics, History—and Change—in the Challenger Explosion. *Washington Post*, January 28th, p.C5.

Webb, J, E. (1963) Address by James Webb, Institute of Foreign Affairs, 24th January p. 2. Available from NASA History Office, Washington DC.

Webb, J, E. (1966) James, E. Webb, Administrator, NASA, to Honourable Everett Dirksen, U.S. Senate, August 9th, in Logsdon, J., Lear, L, J. Warren-Findley, J. Williamson, R, A. and Day, D, W. (Eds.) (1995) *NASA and the Exploration of Space: volume 1* (SP-4218), NASA History Office: Washington DC, pp. 490-2

Webb, J, E. (1969a). Interview with James E. Webb, in Swanson, G, E. (Ed.) *Before This Decade Is Out: Personal Reflections on the Apollo Programe* (SP-4223), NASA History Office: Washington D.C. pp. 1-12

Webb, J, E. (1969b) *Space Age Management: The Large Scale Approach*. McGraw Hill: New York City, NY.

Weitekamp, M. A (2004) *Right Stuff, Wrong Sex: America's First Women*, Johns Hopkins University Press: Baltimore, MD.

Wilford, J, N. (1967) NASA Engineers Criticize Test Schedule Pace, *New York Times,* 8th February.

Williamson, R, A. (1987) Outer Space as Frontier: Lessons for Today, *Western Folklore*, 46 (October), 255-67.

Wolfe, T. (1979) *The Right Stuff,* Black Dog & Leventhal Publishers: New York City, NY.

Woods, B. (2009) A political history of NASA's space shuttle: the development years, 1972-1982, in Bell, D. and Parker, M. (Eds) *Space Travel & Culture: From Apollo to Space Tourism*, Wiley-Blackwell: Oxford, pp. 25-46.

Young, K. (2002) Service pays tribute to crew of Apollo 1, *Florida Today*, January 28th.

Index